Lecture Notes in Mathematics 1579

Editors:
A. Dold, Heidelberg
B. Eckmann, Zürich
F. Takens, Groningen

T0222700

Norihiko Kazamaki

Continuous Exponential Martingales and *BMO*

Springer-Verlag

Berlin Heidelberg New York
London Paris Tokyo
Hong Kong Barcelona
Budapest

Author

Norihiko Kazamaki
Department of Mathematics
Faculty of Science
Toyama University
Gofuku, Toyama 930, Japan

Mathematics Subject Classification (1991): 60G

ISBN 3-540-58042-5 Springer-Verlag Berlin Heidelberg New York
ISBN 0-387-58042-5 Springer-Verlag New York Berlin Heidelberg

CIP-Data applied for

© Springer-Verlag Berlin Heidelberg 1994
Printed in Germany

Typesetting: Camera-ready by author/editor
SPIN: 10130035   46/3140-543210 - Printed on acid-free paper

Preface

This book consists of three chapters and we shall deal entirely with continuous local martingales. Let M be a continuous local martingale and let

$$\mathcal{E}(M) = \exp\left(M - \frac{1}{2}\langle M \rangle\right)$$

where $\langle M \rangle$ denotes the increasing process associated with M. As is well-known, it is a local martingale which plays an essential role in various questions concerning the absolute continuity of probability laws of stochastic processes. Our purpose here is to make a full report on the exciting results about BMO in the theory of exponential local martingales. BMO denotes the class of all uniformly integrable martingales $M = (M_t, \mathcal{F}_t)$ such that

$$\sup_T \left\| E\left[|M_\infty - M_T| \mid \mathcal{F}_T \right] \right\|_\infty < \infty$$

where the supremum is taken over all stopping times T. A martingale in BMO is a probabilistic version of a function of bounded mean oscillation introduced in [31] by F. John and L. Nirenberg.

In Chapter 1 we shall explain in detail the beautiful properties of an exponential local martingale. In Chapter 2 we shall collect the main tools to study various properties about continuous BMO-martingales. The fundamentally important result is that the following are equivalent:

(a) $M \in BMO$.

(b) $\mathcal{E}(M)$ is a uniformly integrable martingale which satisfies the reverse Hölder inequality :

(R_p) $E[\mathcal{E}(M)_\infty^p \mid \mathcal{F}_T] \leq C_p \mathcal{E}(M)_T^p$

for some $p > 1$, where T is an arbitrary stopping time.

(c) $\mathcal{E}(M)$ satisfies the condition:

(A_p) $\sup_T \left\| E\left[\{\mathcal{E}(M)_T / \mathcal{E}(M)_\infty\}^{\frac{1}{p-1}} \mid \mathcal{F}_T \right] \right\|_\infty < \infty$

for some $p > 1$.

These three conditions were originally introduced in the classical analysis. For example, the (A_p) condition is a probabilistic version of the one introduced in [62] by B. Muckenhoupt. In Chapter 3 we shall prove that it is a necessary and sufficient condition for the validity of some weighted norm inequalities for martingales. Furthermore, we shall study two important subclasses of BMO, namely, the class L_∞ of all bounded martingales and the class H_∞ of all martingales M such that $\langle M \rangle_\infty$ is bounded. In general, BMO is neither L_∞ nor H_∞ and it is obvious that there is no inclusion relation between L_∞ and H_∞. In this chapter we shall establish very interesting relationships between the condition (R_p) and the distance to L_∞ in the space BMO. One of them is the result that M belongs to the BMO-closure of L_∞ if and only if $\mathcal{E}(\lambda M)$ satisfies all (R_p) for every real number λ. In addition, we shall prove that the (A_p) condition is remotely related to the distance to H_∞ in the space BMO.

The reader is assumed to be familiar with the martingale theory as expounded in [12] or [60].

I am happy to acknowledge the influence of three of my teachers T. Tsuchikura, C. Watari, and P. A. Meyer. I would also like to thank my colleagues M. Izumisawa, M. Kaneko, M. Kikuchi, M. Okada, T. Okada, T. Sekiguchi, and Y. Shiota for many helpful discussions. Finally, thanks to Mrs. Yoshiko Kitsunezuka for the help in preparing this manuscript.

N. Kazamaki

Contents

Chapter 1

Exponential Martingales

1.1 Preliminaries

Let $(\Omega, \mathcal{F}, P)$ be a complete probability space with a filtration $(\mathcal{F}_t)$ satisfying the usual conditions. The usual hypotheses means that

(i) $\mathcal{F}_0$ contains all the P-null sets of $\mathcal{F}$,
(ii) $\mathcal{F}_t = \bigcap_{u>t} \mathcal{F}_u$ for all $t \geq 0$.

Definition 1. 1.   *A real valued stochastic process $M = (M_t, \mathcal{F}_t)$ is called a* martingale *(resp.supermartingale, submartingale) if*

(i)   *each M_t is $\mathcal{F}_t$-measurable, i.e, M is adapted to the filtration $(\mathcal{F}_t)$,*
(ii)   *$M_t \in L_1$ for every t,*
(iii) *if $s \leq t$, then $E[M_t \mid \mathcal{F}_s] = M_s$ a.s.   (resp. $E[M_t \mid \mathcal{F}_s] \leq M_s$, resp.$\geq M_s$).*

As is well known, a one-dimensional Brownian motion is a typical martingale. It should be noted that the notion of a martingale depends merely on the filtration $(\mathcal{F}_t)$, but also on the probability measure dP. An adapted process $M = (M_t, \mathcal{F}_t)$ is said to be *a local martingale* if there exists a sequence of increasing stopping times T_n with $\lim_{n \to \infty} T_n = \infty$ a.s. such that $(M_{t \wedge T_n} I_{\{T_n > 0\}}, \mathcal{F}_t)$ is a martingale for each n. Such a sequence (T_n) of stopping times is called *a fundamental sequence.* Recall that a stopping time T is a random variable taking values in $[0, \infty]$ such that $\{T \leq t\} \in \mathcal{F}_t$ for every $t \geq 0$.
Throughout this survey, we suppose that any local martingale adapted to this filtration is continuous. It is well-known that the Brownian filtration satisfies this assumption. Note that the following three properties are equivalent :

(a)   any local martingale is continuous,
(b)   any stopping time is predictable,
(c)   for every stopping time T and every $\mathcal{F}_T$-measurable random variable
      U, there exists a continuous local martingale M with $M_T = U$ a.s.

The equivalence of (a) and (b) is well-known, and the equivalence of (a) and (c) was established by M. Emery, C. Stricker and J. A. Yan ([16]). It seems to me that the essential feature of our problems discussed here appears in this case, which is the reason that we deal entirely with continuous local martingales. We generally assume that $M_0 = 0$. Let us denote by $\langle M \rangle$ the continuous increasing process such that $M^2 - \langle M \rangle$ is also a local martingale. Let $t > 0$ and let $\{T_i^n\}_{i=0,1,\cdots,k_n}$ be a sequence of stopping times such that $0 = T_0^n \leq T_1^n \leq \cdots \leq T_{k_n}^n = t$ and $\lim_{n\to\infty} \max_{0 \leq i < k_n}(T_{i+1}^n - T_i^n) = 0$. Then from a celebrated result of C. Doléans-Dade([8]) it follows that

$$\sum_{i=0}^{k_n-1} (M_{T_{i+1}^n} - M_{T_i^n})^2 \longrightarrow \langle M \rangle_t$$

in probability as $n \to \infty$. An adapted process $X = (X_t, \mathcal{F}_t)$ is said to be a *semimartingale* if X_t can be written as $M_t + A_t$ where M is a local martingale and A is a stochastic process that is locally of bounded variation. Let $\langle X \rangle = \langle M \rangle$ as usual.

The next formula plays an extremaly important role in stochastic calculus.

Theorem 1. 1. (Itô's formula)
Let $X = M + A$ be a continuous semimartingale, and let f be a real valued function on R which is twice continuously differentiable. Then

$$(1.1) \qquad f(X_t) = f(X_0) + \int_0^t f'(X_s)dX_s + \frac{1}{2}\int_0^t f''(X_s)d\langle M \rangle_s.$$

Note that the second term on the right hand side is the stochastic integral.

Proof.   We shall sketch its proof. Let $f \in C^2$. The proof rests essentially on Taylor's theorem :

$$(1.2) \qquad f(y) - f(x) = f'(x)(y - x) + \frac{1}{2}f''(x)(y - x)^2 + R(x, y)$$

where $|R(x,y)| \leq r(|y - x|)(y - x)^2$, such that $r : \mathbb{R}_+ \mapsto \mathbb{R}_+$ is an increasing function satisfying $\lim_{u\downarrow 0} r(u) = 0$. Let now X be a continuous semimartingale. Without loss of generality we can take $X_0 = 0$, and further by stopping at $T_m = \inf\{t : |X_t| \geq m\}$, we may assume that X is bounded. Let t be a fixed positive number, and let $\{T_i^n\}_{i=0,1,\cdots,k_n}$ be a sequence of stopping times such that $0 = T_0^n \leq T_1^n \leq \cdots \leq T_{k_n}^n = t$ and $\lim_{n\to\infty} \max_{0 \leq i < k_n}(T_{i+1}^n - T_i^n) = 0$. Then it is not difficult to see that

$$\sum_{i=0}^{k_n-1} (X_{T_{i+1}^n} - X_{T_i^n})^2 \longrightarrow \langle X \rangle_t$$

in probability as $n \to \infty$. From (1.2) it follows that

$$f(X_t) - f(X_0)$$
$$= \sum_{i=0}^{k_n-1} \left\{ f\left(X_{T_{i+1}^n}\right) - f\left(X_{T_i^n}\right) \right\}$$
$$= \sum_{i=0}^{k_n-1} f'\left(X_{T_i^n}\right)\left(X_{T_{i+1}^n} - X_{T_i^n}\right)$$
$$+ \frac{1}{2}\sum_{i=0}^{k_n-1} f''\left(X_{T_i^n}\right)\left(X_{T_{i+1}^n} - X_{T_i^n}\right)^2 + \sum_{i=0}^{k_n-1} R\left(X_{T_i^n}, X_{T_{i+1}^n}\right)$$

The first sum in the last expression converges in probability to the stochastic integral $\int_0^t f'(X_s)dX_s$, and the second sum converges in probability to $\frac{1}{2}\int_0^t f''(X_s)d\langle X \rangle_s$. So it remains to consider the third sum. To estimate it, observe that

$$\lim_{n\to\infty} \max_i r\left(\left|X_{T_{i+1}^n} - X_{T_i^n}\right|\right) = 0,$$

which follows from the assumption that X is bounded and continuous. Then

$$\left| \sum_{i=0}^{k_n-1} R\left(X_{T_i^n}, X_{T_{i+1}^n}\right) \right| \leq \max_{0 \leq i < k_n} r\left(\left|X_{T_{i+1}^n} - X_{T_i^n}\right|\right) \sum_{i=0}^{k_n-1} (X_{T_{i+1}^n} - X_{T_i^n})^2,$$

and the right-hand side converges in probability to 0 as $n \to \infty$. Thus (1.1) holds. $\square$

The Ito formula shows that the class of semimartingales is invariant under composition with C^2- function.

Theorem 1. 2. *If M is a continuous local martingale, then*

$$(1.3) \qquad \mathcal{E}(M)_t = \exp\left(M_t - \frac{1}{2}\langle M \rangle_t\right) \quad (0 \leq t < \infty)$$

is also a local martingale such that $\mathcal{E}(M)_0 = 1$.

Proof. Applying Itô's formula with $X = M - \frac{1}{2}\langle M \rangle$ and $f(x) = e^x$ we obtain

$$\mathcal{E}(M)_t = 1 + \int_0^t \mathcal{E}(M)_s dM_s,$$

which completes the proof. $\square$

In [55] B. Maisonneuve gave a nice proof without using Ito's formula, which will be presented at the end of this section.

The generalization to non-continuous local martingales was done in 1970 by C. Doléans-Dade ([9]). She proved that if X is a semimartingale with $X_{0-} = 0$, then the solution Y of the stochastic integral equation

$$Y_t = 1 + \int_0^t Y_{s-} dX_s$$

is given by the formula

$$Y_t = \exp\left(X_t - \frac{1}{2}\langle X^c \rangle_t\right) \prod_{0 \leq s \leq t} (1 + \triangle X_s) e^{-\triangle X_s},$$

where $\triangle X_s = X_s - X_{s-}$ and X^c is the continuous part of X.

A noteworthy fact is that, *supposing that the exponential local martingale $\mathcal{E}(M)$ is uniformly integrable, it is not necessarily a true martingale.* We first give such an example.

Example 1.1. Let $B = (B_t, \mathcal{F}_t)$ be a 3-dimensional Brownian motion starting at a $(a \neq o)$, and for $0 < r < |a|$ let $\tau_r = \inf\{t : |B_t| \leq r\}$. Then it is well-known that $P(\tau_r < \infty) = r/|a|$.
Let now h be the function defined by $h(x) = |a|/|x|$ for $x \in \mathbb{R}^3 \backslash \{o\}$ which is obviously superharmonic in $\mathbb{R}^3$ and harmonic in the domain $\{x \in \mathbb{R}^3 : |x| > r\}$. Then the process Z defined by $Z_t = h(B_t)$ $(0 \leq t < \infty)$ is a positive supermartingale such that $Z_0 = 1$. By Doob's convergence theorem Z_t converges almost surely and in L_1 as $t \to \infty$. Then, the family $\{Z_t\}_{0 \leq t < \infty}$ being compact in L_1, Z is uniformly

integrable. Next, let $T_n = \tau_{1/n}$ $(n = 1, 2, \cdots)$. It is clear that $T_n \uparrow \infty$ a.s. Moreover, $Z^{T_n} = h(B^{T_n})$ is a martingale, because h is harmonic in $\{x \in \mathbb{R}^3 : |x| > 1/n\}$. Namely, Z is a local martingale which is uniformly integrable. However, it is impossible that Z is a martingale. Observe that $Z = \mathcal{E}(M)$ where $M_t = \int_0^t Z_s^{-1} dZ_s$ $(0 \leq t < \infty)$.

Generally, we have $E[\mathcal{E}(M)_t] \leq 1$ for every t, because $\mathcal{E}(M)$ is a positive super-martingale with $\mathcal{E}(M)_0 = 1$. Therefore, it is a martingale if and only if $E[\mathcal{E}(M)_t] = 1$ for every t. But the direct verification is usually hard to carry out. We shall deal the problem of finding sufficient conditions for $\mathcal{E}(M)$ to be a martingale in Sections 2 and 4.

An easy calculation shows that

$$\mathcal{E}(M)\mathcal{E}(-M) = \exp(-\langle M \rangle), \quad \mathcal{E}(M) = \mathcal{E}\left(\frac{1}{2}M\right)^2 \exp\left(-\frac{1}{4}\langle M \rangle\right)$$

From these relations one can immediately derive that

$$(1.4) \qquad\qquad \{\mathcal{E}(M)_\infty = 0\} = \{\langle M \rangle_\infty = \infty\}.$$

Example 1.2.   Let $B = (B_t, \mathcal{F}_t)$ be a one dimensional Brownian motion starting at 0. For each $t > 0$ we have

$$\begin{aligned} E[\mathcal{E}(B)_t] &= \int_{-\infty}^{\infty} \exp\left(x - \frac{t}{2}\right) \frac{1}{\sqrt{2\pi t}} \exp\left(-\frac{x^2}{2t}\right) dx \\ &= \int_{-\infty}^{\infty} \frac{1}{\sqrt{2\pi t}} \exp\left(-\frac{(x-t)^2}{2t}\right) dx \\ &= 1 \end{aligned}$$

and hence $\mathcal{E}(B)$ is a *true* martingale. However, since $\mathcal{E}(B)_\infty = 0$ a.s., it is not a uniformly integrable martingale.

The martingale property of $\mathcal{E}(B)$ is used effectively in the book of Mckean ([56]) for computing the distribution of quantities associated with Brownian motion. In the following we give such an examples.

Example 1.3.   Let $\tau = \inf\{t : B_t = at + b\}$ where $a \in R$ and $b > 0$. Then we have

$$(1.5) \qquad\qquad P(\tau < \infty) = \exp(-2a^+ b).$$

To see this, observe that $\mathcal{E}(\alpha B^\tau) \leq \exp(\alpha b)$. This implies that $\mathcal{E}(\alpha B^\tau)$ is a uniformly integrable martingale. Then

$$1 = E[\mathcal{E}(\alpha B^\tau)_\infty] = E\left[\exp\left(a\alpha\tau + b\alpha - \alpha^2 \frac{\tau}{2}\right) : \tau < \infty\right].$$

If $a > 0$, setting $\alpha = a + \sqrt{a^2 + 2\lambda}$ we have

$$E[\exp(ab + b\sqrt{a^2 + 2\lambda}) \exp(-\lambda\tau) : \tau < \infty] = 1,$$

that is, $E[\exp(-\lambda\tau) : \tau < \infty] = \exp(-ab - b\sqrt{a^2 + 2\lambda})$. Letting $\lambda \to 0$ we obtain (1.5).

Theorem 1. 3. (D. W. Stroock and S. R. S. Varadhan [80])
Let $M = (M_t, \mathcal{F}_t)$ be a continuous process and let $A = (A_t, \mathcal{F}_t)$ be a continuous process of finite variation such that $A_0 = 0$. Suppose that for sufficiently small λ the process $Z^{(\lambda)}$ defined by

$$Z_t^{(\lambda)} = \exp\left(\lambda M_t - \frac{1}{2}\lambda^2 A_t\right)$$

is a local martingale. Then M is a local martingale with $A = \langle M \rangle$.

Proof. We sketch the proof. By the assumption there is a $\lambda_0 > 0$ such that for any λ with $|\lambda| \leq \lambda_0$ $Z^{(\lambda)}$ is a local martingale. Let now $0 \leq s < t$. The usual stopping argumment enables us to assume that both $\exp(\lambda_0 M_t^*)$ and A_t are integrable where $M_t^* = \sup_{0 \leq s \leq t} |M_s|$. Then for every λ with $|\lambda| \leq \lambda_0$ $Z^{(\lambda)}$ is a martingale, so that for every $D \in \mathcal{F}_s$

$$\int_D Z_t^{(\lambda)} dP = \int_D Z_s^{(\lambda)} dP.$$

Differentiating the both sides with respect to λ we have

(1.6) $$\int_D (M_t - \lambda A_t) Z_t^{(\lambda)} dP = \int_D (M_s - \lambda A_s) Z_s^{(\lambda)} dP.$$

Noticing $Z_0^{(\lambda)} = 1$ and setting $\lambda = 0$ gives

$$\int_D M_t dP = \int_D M_s dP \quad (D \in \mathcal{F}_s).$$

This implies that M is a martingale. Further, taking again the derivatives of the both sides in (1.6) with respect to λ at $\lambda = 0$, we find that $M^2 - A$ is a martingale, that is, $A = \langle M \rangle$. □

Remark 1.1. The continuity of the process A and the condition $A_0 = 0$ are essential for the validity of this theorem (see J. Stoyanov [78],p.256).

The extension of Ito's formula to functions of several semimartingales is the following.

Theorem 1. 4. *If $X = (X^1, X^2, \cdots, X^n)$ is an n-tuple of continuous semimartingales and $f : \mathbb{R}^n \to \mathbb{R}$ has continuous second-order partial derivatives, then*

$$f(X_t) - f(X_0)$$
$$= \sum_{i=1}^n \int_0^t D_i f(X_s) dX_s^i + \frac{1}{2} \sum_{1 \leq i,j \leq n} \int_0^t D_{ij} f(X_s) d\langle X^i, X^j \rangle_s.$$

For the proof, see [12].

A supplementary note. We close this section with another proof of Theorem 1.2 given in [55] by B. Maisonneuve. It is based on the next two lemmas.

Lemma 1. 1. *Let $f = (f_n, \mathcal{G}_n)_{n=0,1,2,\cdots}$ be a martingale, and let*

$$g_n = \frac{e^{f_n}}{\prod_{i=1}^n E[e^{\Delta f_i} | \mathcal{G}_{i-1}]}$$

where $\Delta f_i = f_i - f_{i-1}$ $(i = 1, 2, \cdots)$. Then $g = (g_n, \mathcal{G}_n)$ is a martingale.

This follows immediately by an elementary calculation.

Lemma 1. 2.   *Let $M = (M_t, \mathcal{F}_t)_{0 \leq t < \infty}$ be a continuous local martingale. Then for each $t > 0$ there exist partitions $\Gamma_n : 0 = t_0^n < t_1^n < \cdots < t_{m_n}^n = t$ of $[0, t]$ such that $\lim_{n \to \infty} \max_{1 \leq i \leq m_n}(t_i^n - t_{i-1}^n) = 0$ and*

$$\prod_{i=1}^{m_n} E\left[e^{\triangle M_i^n} | \mathcal{F}_{t_{i-1}^n}\right] \longrightarrow \exp(\frac{1}{2}\langle M \rangle_t) \quad a.s \quad (n \to \infty)$$

where $\triangle M_{t_i^n} = M_{t_i^n} - M_{t_{i-1}^n} \quad (i = 1, 2, \cdots, m_n)$.

Proof.   Let M be a continuous local martingale. The usual stopping time argument enables us to assume that $|M| \leq K$ for some constant $K > 0$.
Let now $h(x) = e^x - 1 - x$ and $k(x) = \log(1 + x) - x$. Then

(1.7)
$$0 \leq h(x) \leq \frac{x^2}{2} e^{|x|} \quad (x \in \mathbb{R})$$

(1.8)
$$\left| h(x) - \frac{x^2}{2} \right| \leq C|x|^3 \quad (|x| \leq 2K)$$

(1.9)
$$-\frac{x^2}{2} \leq k(x) \leq 0 \quad (0 \leq x < \infty),$$

where the constant C depends only on K.
For a partition $\Gamma : 0 = t_0 < t_1 < \cdots < t_{m-1} < t_m = t$ of $[0, t]$, we set

$$\begin{aligned}
\|\Gamma\| &= \max_{1 \leq i \leq m}(t_i - t_{i-1}) \\
U_i &= E[h(\triangle M_{t_i})|\mathcal{F}_{t_{i-1}}], \\
W_i &= E\left[h(\triangle M_{t_i}) - \frac{1}{2}\triangle M_{t_i}^2 \Big| \mathcal{F}_{t_{i-1}}\right], \\
\mathcal{P}_\Gamma &= \prod_{i=1}^m E[\exp(\triangle M_{t_i})|\mathcal{F}_{t_{i-1}}].
\end{aligned}$$

Then, combining (1.7), (1.8) and (1.9) with the sample continuity of M shows that

$$\begin{aligned}
E\left[\sum_{i=1}^m |W_i|\right] &\leq CE\left[\sum_{i=1}^m |\triangle M_{t_i}|^3\right] \\
&\leq CE\left[\sup_{1 \leq i \leq m} |\triangle M_{t_i}| \sum_{j=1}^m \triangle M_{t_j}^2\right] \longrightarrow 0 \quad (\|\Gamma\| \to 0),
\end{aligned}$$

$$\begin{aligned}
E[\sum_{i=1}^m |k(U_i)|] &\leq \frac{1}{2}\sum_{i=1}^m E[E[h(U_i)|\mathcal{F}_{t_{i-1}}]^2] \\
&\leq \frac{1}{2}\sum_{i=1}^m E\left[E\left[\frac{\triangle M_{t_i}^2}{2}\exp(|\triangle M_{t_i}|)\Big|\mathcal{F}_{t_{i-1}}\right]^2\right] \\
&\leq \frac{4K}{8}E\left[\sum_{i=1}^m \triangle M_{t_i}^4\right] \\
&\leq \frac{4K}{8}E\left[\sup_{1 \leq j \leq m} \triangle M_{t_j}^2 \sum_{i=1}^m \triangle M_{t_i}^2\right] \longrightarrow 0 \quad (n \to \infty).
\end{aligned}$$

On the other hand, since $E[\exp(\triangle M_{t_i})|\mathcal{F}_{t_{i-1}}] = 1 + U_i$, we have

$$
\begin{aligned}
\mathcal{P}_\Gamma &= \prod_{i=1}^{m}(1+U_i) \\
&= \exp\left\{ \sum_{i=1}^{m} U_i + \sum_{i=1}^{m} k(U_i) \right\} \\
&= \exp\left\{ \frac{1}{2}\sum_{i=1}^{m} E\left[\triangle M_{t_i}^2|\mathcal{F}_{t_{i-1}}\right] + \sum_{i=1}^{m} W_i + \sum_{i=1}^{m} k(U_i) \right\}.
\end{aligned}
$$

Recall that

$$
\sum_{i=1}^{m} E\left[\triangle M_{t_i}^2|\mathcal{F}_{t_{i-1}}\right] \longrightarrow \langle M \rangle_t \quad \text{in } L_1 \text{ as } \|\Gamma\| \to 0
$$

by the well-known result of C. Doléans-Dade. Then there exists a sequence of partitions $\Gamma_n : 0 = t_0^n < t_1^n < \cdots < t_{m_n}^n = t$ of $[0,t]$ such that

$$
\mathcal{P}_{\Gamma_n} \longrightarrow \exp\left(\frac{1}{2}\langle M \rangle_t\right) \quad a.s \quad (n \to \infty).
$$

Thus the lemma is proved. □

Now, let $0 < s < t$. Without loss of generality we may assume that $s \in \Gamma_n$ for every n, namely, $s = t_{k_n}^n$ for some $k_n \leq m_n$. Let Γ_n' be the partition $:0 = t_0^n < t_1^n < \cdots < t_{k_n}^n = s$ of $[0,s]$ and let

$$
\mathcal{Q}_{\Gamma_n'} = \prod_{i=1}^{k_n} E\left[\exp(\triangle M_{t_i^n})|\mathcal{F}_{t_{i-1}^n}\right].
$$

Then $\mathcal{Q}_{\Gamma_n'} \longrightarrow \exp(\frac{1}{2}\langle M \rangle_s)$ $a.s$ $(n \to \infty)$ as is already stated above, and from Lemma 1.1 it follows that

$$
E\left[\frac{e^{M_t}}{\mathcal{P}_{\Gamma_n}}\middle|\mathcal{F}_s\right] = \frac{e^{M_s}}{\mathcal{Q}_{\Gamma_n'}}.
$$

Thus, letting $n \to \infty$ we obtain

$$
E\left[\frac{e^{M_t}}{e^{\frac{1}{2}\langle M \rangle_t}}\middle|\mathcal{F}_s\right] = \frac{e^{M_s}}{e^{\frac{1}{2}\langle M \rangle_s}},
$$

which completes the proof.

1.2 The L_p-integrability of $\mathcal{E}(M)$

As is well-known, exponential martingales play an essential role in various questions concerning the absolute continuity of probability laws of stochastic processes. However, $\mathcal{E}(M)$ is not always a uniformly integrable martingales as stated before, and it is often difficult to verify the uniform integrability of $\mathcal{E}(M)$. In 1960, I.V. Girsanov([23]) showed that if $\langle M \rangle_\infty$ is bounded, then $\mathcal{E}(M)$ is a uniformly integrable martingale. In 1972, this assertion was proved by I.I. Gihman and A.V. Skorohod([22]) when $\exp((1+\delta)\langle M \rangle_\infty) \in L_1$ for some $\delta > 0$ and then by R.S. Lipster and A.N. Shiryayev

([54]) when $\exp((\frac{1}{2}+\delta)\langle M\rangle_\infty) \in L_1$ for some $\delta > 0$. After that, A.A. Novikov([64]) gave a nice criterion. In this section we improve his result.

We first prove the following result, from which a simple criterion can be derived. It is remarkable that the constant $\frac{\sqrt{p}}{2(\sqrt{p}-1)}$ is the best possible, and then its proof is extremely simple.

Theorem 1. 5.   *Let $1 < p < \infty$ and $p^{-1} + q^{-1} = 1$. Suppose that*

$$\sup_T E\left[\exp\left(\frac{\sqrt{p}}{2(\sqrt{p}-1)}M_T\right)\right] < \infty,$$

where the supremum is taken over all bounded stopping times T. Then $\mathcal{E}(M)$ is an L_q-bounded martingale.

Proof.   For $1 < p < \infty$, let $r = (\sqrt{p}+1)/(\sqrt{p}-1)$. Then the exponent conjugate to r is $s = (\sqrt{p}+1)/2$, and note that $(q - \sqrt{q/r})s = \sqrt{p}/\{2(\sqrt{p}-1)\}$ by a simple calculation. Since we have

$$\mathcal{E}(M)^q = \exp\left(\sqrt{\frac{q}{r}}M - \frac{q}{2}\langle M\rangle\right)\exp\left\{\left(q - \sqrt{\frac{q}{r}}\right)M\right\},$$

an application of Hölder's inequality shows that for any stopping time S

$$E[\mathcal{E}(M)_S^q] \leq E[\mathcal{E}(\sqrt{qr}M)_S]^{\frac{1}{r}} E\left[\exp\left\{\left(q - \sqrt{\frac{q}{r}}\right)sM_S\right\}\right]^{\frac{1}{s}}$$

$$\leq \sup_T E\left[\exp\left\{\frac{\sqrt{p}}{2(\sqrt{p}-1)}M_T\right\}\right]^{\frac{1}{s}} < \infty.$$

This completes the proof.   □

Remark 1.2.   Let M be a right continuous local martingale such that $\triangle M \geq 0$ and suppose that $E[\exp(\frac{K}{2}[M]_\infty)] < \infty$ for some $K > 0$. Then J. Yan proved in ([86]) that $\mathcal{E}(M) \in H_r$ with $r = \frac{K}{2\sqrt{K}-1}$ if $1 < K \leq 4$ and $r = \frac{2K}{K+2}$ if $K > 4$. In [52] D. Lépingle and J. Mémin improved his result.

Remark 1.3.   H. Sato gave in [71] the following interesting result : Let M be a stochastically continuous additive process with paths which are right continuous and have left-hand limits at every point. If $M_0 = 0, E[M_t] = 0$ and $\triangle M_t > -1$ for every t, then all of the following statements are equivalent.

(i) $M^* \in L_1$
(ii) M is uniformly integrable.
(iii) M_t converges almost surely as $t \to \infty$.
(iv) $\mathcal{E}(M)_\infty > 0$.
(v) $\mathcal{E}(M)$ is a uniformly integrable martingale.
(vi) $\mathcal{E}(M)^* \in L_1$

Now, by using Theorem 1.5 we can give a simple but usefull criterion for the uniform integrability of $\mathcal{E}(M)$.

Theorem 1. 6. *Suppose that*

$$\sup_T E\left[\exp\left(\frac{1}{2}M_T\right)\right] < \infty,$$

where the supremum is taken over all bounded stopping times T. Then $\mathcal{E}(M)$ is a uniformly integrable martingale.

Proof. Let $0 < a < 1$ and choose $p > 1$ such that $\sqrt{p}/(\sqrt{p} - 1) < 1/a$. Then by the assumption $\mathcal{E}(aM)$ is an L_q-bounded martingale, and so it is obviously a uniformly integrable martingale. Since $\mathcal{E}(aM) = \mathcal{E}(M)^{a^2} \exp\{a(1 - a)M\}$, by using Hölder's inequality with exponents a^{-2} and $(1 - a^2)^{-1}$ we find

$$
\begin{aligned}
1 = E[\mathcal{E}(aM)_\infty] &\leq E[\mathcal{E}(M)_\infty]^{a^2} E\left[\exp\left(\frac{a}{1+a}M_\infty\right)\right]^{1-a^2} \\
&\leq E[\mathcal{E}(M)_\infty]^{a^2} E\left[\exp\left(\frac{1}{2}M_\infty\right)\right]^{2a(1-a)}
\end{aligned}
$$

The second term on the right hand side converges to 1 as $a \uparrow 1$. Therefore we have $1 \leq E[\mathcal{E}(M)_\infty]$, which completes the proof. $\qquad\square$

Example 1.4. For $0 < a < \infty$, let $\tau_a = \inf\{t > 0 : B_t = a\}$. Let now $\lambda > 0$. Then $\sup_T E\left[\exp\left(\frac{1}{2}B_T^{\tau_a}\right)\right] \leq e^a$ and so $E[\mathcal{E}(\sqrt{2\lambda}B^{\tau_a})_\infty] = 1$ by Theorem 1.6. Then we have

$$E[\exp(-\lambda\tau_a)] = \exp(-a\sqrt{2\lambda}),$$

and, applying the inversion formula for the Laplace transform gives

$$P(\tau_a = t) = \sqrt{2\pi t^3}\,a\exp\left(-\frac{a^2}{2t}\right).$$

Note that the converse of this theorem is not true (see Example 1.13). As a corollary we can obtain the following criterion, because

$$E\left[\exp\left(\frac{1}{2}M_T\right)\right] \leq E\left[\exp\left(\frac{1}{2}\langle M\rangle_T\right)\right]^{\frac{1}{2}}$$

for every stopping time T.

Corollary 1. 1 (A. A. Novikov [64]). *Suppose that*

$$E\left[\exp\left(\frac{1}{2}\langle M\rangle_\infty\right)\right] < \infty.$$

Then $\mathcal{E}(M)$ is a uniformly integrable martingale.

This is known as Novikov's criterion. Note that $1/2$ is the best constant in these criteria. Following the idea of Novikov, we exemplify it below.

Example 1.5. For $0 < a < 1$, let us define the stopping time

$$T = \inf\{t : B_t \leq at - 1\},$$

and set $M = B^T$. Then it is easy to see that $M_\infty - \frac{1}{2}\langle M \rangle_\infty = (a - \frac{1}{2})T - 1$ and so

$$E[\mathcal{E}(M)_\infty] = e^{-1}E\left[\exp\left\{\left(a - \frac{1}{2}\right)T\right\}\right]$$

$$\leq e^{-1}E\left[\exp\left(\frac{a^2}{2}T\right)\right].$$

Since $1 \geq E[\mathcal{E}(aM)_\infty] = E\left[\exp\left(\frac{a^2}{2}T\right)\right]e^{-a}$, we have

$$E\left[\exp\left(\frac{a^2}{2}\langle M \rangle_\infty\right)\right] = E\left[\exp\left(\frac{a^2}{2}T\right)\right] \leq e^a.$$

However, since $E[\mathcal{E}(M)_\infty] \leq e^{a-1} < 1$, $\mathcal{E}(M)$ is not a uniformly integrable martingale.

Next, by giving an example, we illustrate that the constant $\frac{\sqrt{p}}{2(\sqrt{p}-1)}$ in Theorem 1.5 can not replaced by a smaller number.

Example 1.6. Let $B = (B_t, \mathcal{F}_t)_{0 \leq t < \infty}$ be a one dimensional Brownian motion with $B_0 = 0$. We set

$$\tau = \inf\{t \geq 0 : B_t \leq t - 1\},$$

which is obviously a stopping time. Since $0 < \tau < \infty$ and $B_\tau = \tau - 1$, we have $\mathcal{E}(B^\tau)_\infty = \exp(\tau/2 - 1)$, from which it follows at once that $E[\exp(\tau/2)] \leq e$. Then $E[\mathcal{E}(B^\tau)_\infty] = 1$ by Novikov's criterion . Note that $E[\exp(\tau/2)] = e$ in consequence. Clearly, if $0 < a \leq 1$, then $\exp(a^2\tau/2) \in L_1$ and so $\mathcal{E}(aB^\tau)$ is a uniformly integrable martingale. On the other hand, for $b > 1$ we have $\exp(b^2\tau/2) \notin L_1$, because $E[\mathcal{E}(bB^\tau)_\infty] \leq e^{-b}E[\exp(\tau/2)] \leq e^{-b+1} < 1$.
Let now $1 < \lambda < \sqrt{p}/(\sqrt{p} - 1)$. Note that $q(2\lambda - 1)/\lambda^2 > 1$. Let $M = \alpha B^\tau$ where $\alpha = 1/\lambda$, and consider the exponential martingale $\mathcal{E}(M)$. As $0 < \alpha < 1$, it is a uniformly integrable martingale. Since $B_\tau = \tau - 1$ by the definition of τ, we find

$$E\left[\exp\left(\frac{1}{2}\lambda M_\infty\right)\right] = E\left[\exp\left(\frac{1}{2}B_\tau\right)\right]$$

$$= E\left[\exp\left(\frac{1}{2}\tau\right)\right]e^{-\frac{1}{2}}$$

$$= e^{\frac{1}{2}}.$$

However, since $\mathcal{E}(M)_\infty^q = e^{-\alpha q}\exp\{\frac{1}{2}\frac{q(2\lambda-1)}{\lambda^2}\tau\} \notin L_1$, $\mathcal{E}(M)$ is not an L_q-bounded martingale.

Note that Novikov's criterion works not only for $\mathcal{E}(M)$, but for $\mathcal{E}(-M)$, and, incidentally, $\mathcal{E}(-M)$ is not always a uniformly integrable martigale even if $\mathcal{E}(M)$ is bounded.

Example 1.7. Let $\tau = \inf\{t > 0 : B_t \geq 1\}$, which is a stopping time such that $0 < \tau < \infty$. Then $E[\mathcal{E}(-B^\tau)_\infty] < 1$, although $\mathcal{E}(B^\tau)$ is bounded.

We shall close this section with a remarkable fact which tells a distinct difference between the above two criterions.

Theorem 1. 7. *If $\parallel M \parallel_\infty < \pi/2$, then*

$$E\left[\exp\left(\frac{1}{2}\langle M\rangle_\infty\right)\right] < \infty.$$

However, there exists a local martingale M such that $\parallel M \parallel_\infty = \pi/2$ and $\exp(\frac{1}{2}\langle M\rangle_\infty)$ $\notin L_1$.

To show this, we need the next lemma.

Lemma 1. 3. *Let $a, b > 0$ and $\tau = \inf\{t : B_t \notin (-a, b)\}$. Then we have*

$$E\left[\exp\left(\frac{1}{2}\theta^2\tau\right)\right] = \frac{\cos\left(\dfrac{a-b}{2}\theta\right)}{\cos\left(\dfrac{a+b}{2}\theta\right)} \quad \left(0 \le \theta < \frac{\pi}{a+b}\right).$$

Proof. We first show, following the nice idea of D. Lépingle, that the process X given by

$$X_t = \exp\left(\frac{1}{2}\theta^2 t\right)\cos\theta\left(B_t - \frac{b-a}{2}\right) \quad (0 \le t < \infty)$$

is a local martingale. For that, let

$$F(x, y) = \exp\left(\frac{1}{2}\theta^2 x\right)\cos\theta\left(y - \frac{b-a}{2}\right).$$

Then it is clear that $X_t = F(t, B_t)$. Since $F(x, y)$ is twice continuously differentiable, Itô's formula gives

$$X_t = \cos\left(\frac{b-a}{2}\theta\right) - \theta\int_0^t \exp\left(\frac{1}{2}\theta^2 s\right)\sin\theta\left(B_s - \frac{b-a}{2}\right)dB_s.$$

Thus X is a local martingale. On the other hand, from the definition of τ it follows that $|B_\tau - (b-a)/2| = (a+b)/2$ and so

$$\begin{aligned}
X_\tau &= \exp\left(\frac{1}{2}\theta^2\tau\right)\cos\theta\left(B_\tau - \frac{b-a}{2}\right) \\
&= \exp\left(\frac{1}{2}\theta^2\tau\right)\cos\left(\frac{a+b}{2}\theta\right).
\end{aligned}$$

Let now $0 \le \theta < \pi/(a+b)$. Then $|B_{t\wedge\tau} - (b-a)/2|\theta \le (a+b)/2\theta < \pi/2$, and so

$$X_{t\wedge\tau} \ge \exp\left\{\frac{1}{2}\theta^2(t\wedge\tau)\right\}\cos\left(\frac{a+b}{2}\theta\right) > 0.$$

Thus X^τ is a positive supermartingale, from which it follows that

$$\cos\left(\frac{a+b}{2}\theta\right)E\left[\exp\left(\frac{1}{2}\theta^2\tau\right)\right] \le E[X_0] = \cos\left(\frac{a-b}{2}\theta\right),$$

that is,

$$E\left[\exp\left(\frac{1}{2}\theta^2\tau\right)\right] \le \frac{\cos\left(\dfrac{a-b}{2}\theta\right)}{\cos\left(\dfrac{a+b}{2}\theta\right)} < \infty.$$

Combining this result with the definition of X shows that $X_\tau^* \in L_1$, which implies that $E[X_\tau] = E[X_0]$.

Since $X_0 = \cos\left(\dfrac{a-b}{2}\theta\right)$ and $X_\tau = \exp\left(\dfrac{1}{2}\theta^2\tau\right)\cos\left(\dfrac{a+b}{2}\theta\right)$, the conclusion follows at once.   □

Proof of Theorem 1.7 :   It suffices essentially to verify this theorem in the case where $M = B^\zeta$ for a certain stopping time ζ, because any continuous local martingale can be reduced to stopped Brownian motion by means of a continuous change of time.

Firstly, let us assume that $\| M \|_\infty < \dfrac{\pi}{2}$, and consider the stopping time $\tau = \inf\{t :$ $B_t \notin (-d, d)\}$ where $\| M \|_\infty < d < \dfrac{\pi}{2}$. Then, since $\langle M\rangle_\infty = \zeta < \tau$ and $1 < \dfrac{\pi}{2d}$, it follows from the lemma that

$$E\left[\exp\left(\frac{1}{2}\langle M\rangle_\infty\right)\right] \le E\left[\exp\left(\frac{1}{2}\tau\right)\right] \le \frac{1}{\cos d} < \infty.$$

On the other hand, let $\tau = \inf\left\{t : |B_t| = \frac{\pi}{2}\right\}$, and cosider the martingale $M = B^\tau$. Then we find

$$\begin{aligned}
E\left[\exp\left(\frac{1}{2}\tau\right)\right] &= \lim_{\theta\uparrow 1} E\left[\exp\left(\frac{1}{2}\theta^2\tau\right)\right] \\
&= \lim_{\theta\uparrow 1} \frac{1}{\cos\frac{\pi}{2}\theta} = \infty
\end{aligned}$$

by the lemma. Thus the theorem is proved.   □

1.3   Girsanov's formula

In this section we assume that the exponential process $\mathcal{E}(M)$ given by the formula (1.3) is a uniformly integrable martingale. This means that $d\hat{P} = \mathcal{E}(M)_\infty dP$ is also a probability measure on Ω. We denote by $\hat{E}[\,\cdot\,]$ the expectation over Ω with respect to $\hat{P}$. Observe that the martingale property is not invariant under such a change of law. Fortunately, in 1960 a nice key of settling this trouble was given by I. V. Girsanov ([23]). It comes to this that under the absolutely continuous change in probability measure a Brownian motion is transformed into the sum of a Brownian motion and a second process with sample functions which are absolutely continuous with respect to the Lebesgue measure. In 1974 a natural generalization of this result was obtained by H. Van Schuppen and E. Wong ([72])).

Lemma 1. 4.   *Y is a local martingale relative to $\hat{P}$ if and only if $Y\mathcal{E}(M)$ is a local martingale relative to P.*

Proof. Let Y be a local martingale relative to $\hat{P}$. Without loss of generality we may assume that Y is a martingale. Then for all $s < t$ and $A \in \mathcal{F}_s$,

$$\int_A Y_t d\hat{P} = \int_A Y_s d\hat{P}.$$

Since $d\hat{P} = \mathcal{E}(M)_T dP$ on $\mathcal{F}_T$ for each stopping time T, we have

$$\int_A Y_t \mathcal{E}(M)_t dP = \int_A Y_s \mathcal{E}(M)_s dP.$$

Namely, $Y\mathcal{E}(M)$ is a martingale relative to P.
The converse follows immediatelly from this argument. □

In what follows let $\mathcal{L}$ (resp. $\hat{\mathcal{L}}$) denote the class of all continuous local martingales relative to P (resp. $\hat{P}$).

Theorem 1. 8. *For any $X \in \mathcal{L}$, the process $\hat{X} = \langle X, M \rangle - X$ belongs to the class $\hat{\mathcal{L}}$ and $\langle \hat{X} \rangle = \langle X \rangle$ under either probability measure. Furthermore, the mapping $\phi : X \mapsto \hat{X}$ is linear and bijective.*

The mapping $\phi : \mathcal{L} \to \hat{\mathcal{L}}$ is called the *Girsanov transformation*. It is also often called *transformation of drift.*

Proof. By Lemma 1.4, in order to see $\hat{X} \in \hat{\mathcal{L}}$ it is enough to verify that $\mathcal{E}(M)\hat{X} \in \mathcal{L}$. Clearly, $\hat{X}$ is a semimartingale with respect to P ,and note that

$$\langle X, M \rangle_t = \int_0^t \mathcal{E}(M)_s^{-1} d\langle X, \mathcal{E}(M) \rangle_s.$$

Then, by Theorem 1.4 we have

$$\begin{aligned}
\mathcal{E}(M)_t \hat{X}_t &= \mathcal{E}(M)_0 \hat{X}_0 + \int_0^t \hat{X}_s d\mathcal{E}(M)_s + \int_0^t \mathcal{E}(M)_s d\hat{X}_s + \langle \mathcal{E}(M), X \rangle_t \\
&= \int_0^t \hat{X}_s d\mathcal{E}(M)_s + \int_0^t \mathcal{E}(M)_s dX_s,
\end{aligned}$$

which belongs to $\mathcal{L}$. Similarly, by Theorem 1.4 we have

$$(\hat{X}_t^2 - \langle X \rangle_t)\mathcal{E}(M)_t = \int_0^t (\hat{X}_s^2 - \langle X \rangle_s)d\mathcal{E}(M)_s - 2 \int_0^t \hat{X}_s \mathcal{E}(M)_s dX_s,$$

which implies that $\hat{X}^2 - \langle X \rangle$ is a local martingale with respect to $\hat{P}$. Then it is clear that $\langle \hat{X} \rangle = \langle X \rangle$. From these facts it follows that the mapping g is linear and injective. So it remains to show the surjectivity. As $\hat{M} = \langle M \rangle - M$ and $\langle \hat{M} \rangle = \langle M \rangle$, we have $\mathcal{E}(\hat{M}) = \mathcal{E}(M)^{-1}$, so that for any $X' \in \hat{\mathcal{L}}$, $X = \langle X', \hat{M} \rangle - X'$ belongs to $\mathcal{L}$. Since $\hat{X} = \langle X, M \rangle - X$ is in $\hat{\mathcal{L}}$, $X' - \hat{X} = \langle X', \hat{M} \rangle - \langle X, M \rangle$ is also a $\hat{P}$-continuous local martingale with finite variation on each finite interval. This implies that $X' = \hat{X}$. Thus the theorem is proved. □

J.H.Van Schuppen and E.Wong ([72]) tried to extend this transformation to right continuous local martingales, and in 1977 the generalization was completely established by E. Lenglart ([51]). He proved that the mapping $\phi : \mathcal{L} \to \hat{\mathcal{L}}$ given by

$$\phi(X)_t = X_t - \int_0^t \mathcal{E}(M)_s^{-1} d[X, \mathcal{E}(M)]_s$$

is also well-defined and linear in the general setting.

Note that *the stochastic integral $H \circ \hat{X}$ relative to $\hat{P}$* coincides with *the stochastic integral of H with respect to the semimartingale $\hat{X}$ relative to P.*
We now state a typical application of Theorem 1.8.

Corollary 1. 2. (I. V. Girsanov [23])
Let $B = (B_t, \mathcal{F}_t)$ be a Brownian motion under the measure P and let $H = (H_t, \mathcal{F}_t)$ be a predictable process such that for every $t > 0$

$$\int_0^t H_s^2 ds < \infty \quad a.s.$$

Assume that $\mathcal{E}(H \circ B)_t = \exp(\int_0^t H_s dB_s - \frac{1}{2}\int_0^t H_s^2 ds)$ is a uniformly integrable martingale. Then the process $\hat{B}$ defined by $\hat{B}_t = \int_0^t H_s ds - B_t$ is a Brownian motion under $\hat{P}$.

Proof.   By Theorem 1.8 the process $\hat{B}$ is a continuous local martingale under $\hat{P}$ such that $\langle \hat{B} \rangle_t = t$ for every t. Thus it follows from Lévy's characterization of Brownian motion that $\hat{B}$ is a Brownian motion under $\hat{P}$.   □

For $1 \leq p < \infty$, let H_p denote the class of all local martingales X over $(\mathcal{F}_t)$ such that $\| X \|_{H_p} = E[\langle X \rangle_\infty^{\frac{p}{2}}]^{\frac{1}{p}} < \infty$. If $1 < p < \infty$, then H_p coincides with the class of all L_p-bounded martingales. On the other hand, H_1 coincides with the class of all martingales X such that $X^* = \sup_{0 \leq t < \infty} |X_t| \in L_1$. The next inequality due to B. Davis is well known :

$$(1.10) \qquad \frac{1}{2} E\left[\langle X \rangle_\infty^{\frac{1}{2}}\right] \leq E[X^*] \leq 4\sqrt{2} E\left[\langle X \rangle_\infty^{\frac{1}{2}}\right].$$

Therefore, $X \in H_1$ if and only if $X^* \in L_1$. Each H_p is a real Banach space with the norm $\| \cdot \|_{H_p}$. We shall denote by $\hat{H}_p$ the H_p class associated with $\hat{P}$. Then it is a question whether or not $\hat{X} \in \hat{H}_p$ for $X \in H_p$. The next example shows that the answer is negative.

Example 1.8.   Let $B = (B_t, \mathcal{F}_t)$ be a one-dimensional Brownian motion starting at 0 where $(\mathcal{F}_t)$ is the filtration generated by (B_t). Recall that any local martingale adapted to this filtration is continuous.
Let now $\tau = \inf\{t : |B_t| = 1\}$, and consider the martingale $X = B^\tau$. Then $|X| \leq 1$, but $\tau^{1/2} \notin L_\infty$ clearly by the definition of τ. Since the dual of L_1 is L_∞, there exists a random variable $Z > 0$ *a.s* such that $E[Z] = 1$ and $E[Z\langle X \rangle_\infty^{\frac{1}{2}}] = \infty$. Next, we define the two continuous martingales $Z_t = E[Z|\mathcal{F}_t]$ and $M_t = \int_0^t \frac{1}{Z_s} dZ_s$. It is clear that the martingale Z satisfies the stochastic integral equation $Z_t = 1 + \int_0^t Z_s dM_s$ and so Z is nothing but the exponential $\mathcal{E}(M)$ of M. Then, letting $d\hat{P} = ZdP$, we find that

$$\hat{E}\left[\langle \hat{X} \rangle_\infty^{\frac{1}{2}}\right] = E\left[Z\langle X \rangle_\infty^{\frac{1}{2}}\right] = \infty.$$

Namely, the martingale X is bounded, but $\hat{X}$ does not belong even to the class $\hat{H}_1$.

Theorem 1. 9. $\phi(M) \in H_2(\hat{P})$ if and only if $\mathcal{E}(M) - 1 \in H_1(P)$.

Proof. We begin with the proof of the "if" part. To do it, we need the inequality :

$$E[\mathcal{E}(M)_\infty \log \mathcal{E}(M)_\infty] \le 4\sqrt{2}\pi(E[\mathcal{E}(M)^*] + 1),$$

which follows immediately from a result given by S. Watanabe ([84]). After his idea we show this . Firstly, let us choose Y in $\mathcal{L}$ in such a way that $U_t = \mathcal{E}(M)_t + iY_t$ is a conformal martingale ; that is, $\langle \mathcal{E}(M) \rangle = \langle Y \rangle$ and $\langle \mathcal{E}(M), Y \rangle = 0$. Then $V_t = U_t \log U_t$ is also a conformal martingale, for the function $z \log z$ is analytic in the domain $D = \{z \in C : Rez > 0\}$. Thus, $ReV_t = \mathcal{E}(M)_t \log |U_t| - Y_t \arg U_t$ is a continuous local martingale. By using the stopping time argument we may assume that both $\mathcal{E}(M) \log |U|$ and Y are in $H_2(P)$. Then

$$E[\mathcal{E}(M)_\infty \log |U_\infty|] = E[Y_\infty \arg U_\infty].$$

In addition, $U_\infty \in D$ and so $|\arg U_\infty| \le \pi/2$. An application of the Davis inequality gives

$$
\begin{aligned}
E[\mathcal{E}(M)_\infty \log \mathcal{E}(M)_\infty] &\le E[\mathcal{E}(M)_\infty \log |U_\infty|] \\
&\le \frac{\pi}{2} E[|Y_\infty|] \\
&\le 2\sqrt{2}\pi E[\langle \mathcal{E}(M) \rangle_\infty^{1/2}] \\
&\le 4\sqrt{2}\pi E[(\mathcal{E}(M) - 1)^*].
\end{aligned}
$$

Therefore, if $\mathcal{E}(M) - 1 \in H_1(P)$, then $E[\mathcal{E}(M)_\infty \log \mathcal{E}(M)_\infty] < \infty$. We are now going to show that $\phi(M) \in H_2(\hat{P})$. The stopping time argument enables us to assume that $\hat{M}$ is a uniformly integrable martingale with respect to $\hat{P}$. Then we have

$$
\begin{aligned}
\hat{E}[\langle \hat{M} \rangle_\infty] &= 2\hat{E}\left[-\hat{M}_\infty + \frac{1}{2}\langle \hat{M} \rangle_\infty\right] \\
&= 2E\left[\mathcal{E}(M)_\infty \left(M_\infty - \frac{1}{2}\langle M \rangle_\infty\right)\right] \\
&= 2E[\mathcal{E}(M)_\infty \log \mathcal{E}(M)_\infty].
\end{aligned}
$$

Thus $\hat{M} \in H_2(\hat{P})$.

Now, we show the "only if" part. From the definition of the weighted probability $\hat{P}$ it follows that

$$E[\mathcal{E}(M)_\infty \log^+ \mathcal{E}(M)_\infty] = \hat{E}[\log^+ \mathcal{E}(M)_\infty] = \hat{E}\left[M_\infty - \frac{1}{2}\langle M \rangle_\infty : \mathcal{E}(M)_\infty \ge 1\right].$$

By Theorem 1.8 the right hand side is

$$\hat{E}\left[\hat{M}_\infty + \frac{1}{2}\langle \hat{M} \rangle_\infty : \mathcal{E}(M)_\infty \ge 1\right] \le \hat{E}[\langle \hat{M} \rangle_\infty]^{1/2} + \frac{1}{2}\hat{E}[\langle \hat{M} \rangle_\infty].$$

Therefore, if $\hat{M} \in H_2(\hat{P})$, we have $\mathcal{E}(M)^* \in L_1$ and so $\mathcal{E}(M) - 1 \in H_1(P)$ by the classical inequality of J. L. Doob. Thus the proof is complete. □

Theorem 1. 10. *The spaces H_2 and $\hat{H}_2$ are always isometrically isomorphic under the mapping Φ_2.*

Proof. Let $X \in H_2$. Then $\hat{X}$ is in $\hat{\mathcal{L}}$. Let $T_n \uparrow \infty$ be stopping times such that $\hat{X}^{T_n} \in \hat{H}_2$ for every n. Observing that $\mathcal{E}(\hat{M}) = \mathcal{E}(M)^{-1}$, we have for every $n \geq 1$

$$
\begin{aligned}
\hat{E}[(\mathcal{E}(M)^{-\frac{1}{2}} \circ \hat{X})_{T_n}] &= \hat{E}\left[\int_0^{T_n} \mathcal{E}(\hat{M})_s d\langle \hat{X}\rangle_s\right] \\
&= \hat{E}[\mathcal{E}(\hat{M})_{T_n}\langle \hat{X}\rangle_{T_n}] \\
&= E[\langle X\rangle_{T_n}].
\end{aligned}
$$

Letting $n \to \infty$ and using the monotone convergence theorem we obtain

$$
E[(\mathcal{E}(M)^{-\frac{1}{2}} \circ \hat{X})_\infty] = E[\langle X\rangle_\infty] < \infty,
$$

so that $\mathcal{E}(M)^{-\frac{1}{2}} \circ \hat{X} \in \hat{H}_2$ for $X \in H_2$. This implies that the mapping $\Phi_2 : H_2 \mapsto \hat{H}_2$ given by $\Phi_2(X) = \mathcal{E}(M)^{-\frac{1}{2}} \circ \hat{X}$ is well-defined and further $\| \Phi_2(X) \|_{\hat{H}_2} = \| X \|_{H_2}$. Thus, it remains to prove the surjectivity. To verify this, let $X' \in \hat{H}_2$. By Theorem 1.8 there exists some $Y \in \mathcal{L}$ such that $X' = \phi(Y)$ and $\langle Y\rangle = \langle X'\rangle$. We now set $X = \mathcal{E}(M)^{\frac{1}{2}} \circ Y$ and choose stopping times $T_n \uparrow \infty$ such that $Y^{T_n} \in H_2$ for every n. Then we have

$$
\begin{aligned}
E[\langle X\rangle_{T_n}] &= E\left[\int_0^{T_n} \mathcal{E}(M)_s d\langle Y\rangle_s\right] \\
&= E[\mathcal{E}(M)_{T_n}\langle Y\rangle_{T_n}] \\
&= \hat{E}[\langle X'\rangle_{T_n}] \\
&\leq \hat{E}[\langle X'\rangle_\infty].
\end{aligned}
$$

From Fatou's lemma it follows that $X \in H_2$. Moreover, we have

$$
\mathcal{E}(M)^{-\frac{1}{2}} \circ (\mathcal{E}(M)^{\frac{1}{2}} \circ \hat{Y}) = \hat{Y} = X'.
$$

Consequently, the mapping Φ_2 is surjective. $\square$

In Chapter 3 we shall prove that if M is a BMO-martingale, then the mapping $\Phi_p : X \mapsto \mathcal{E}(M)^{-\frac{1}{p}} \circ \phi(M)$ is an isomorphism of H_p and $\hat{H}_p$.

1.4   Uniform integrability of $\mathcal{E}(M)$

We have given in Section 2 two sufficient conditions for the uniform integrability of an exponential martingale. The purpose of this section is to improve these criterions. From now on let $\mathcal{M}_u$ denote the class of all uniformly integrable martingales for convenience.

Lemma 1. 5. *Let M be a continuous local martingale and for $\lambda > 0$ let*

$$
T_\lambda = \inf\{t \geq 0 : \langle M\rangle_t > \lambda\}.
$$

Then $\mathcal{E}(M) \in \mathcal{M}_u$ if and only if $\liminf_{\lambda \to \infty} E[\mathcal{E}(M)_{T_\lambda} : T_\lambda < \infty] = 0$.

Proof. For each $\lambda > 0$ we find that $\langle M \rangle_{T_\lambda} \leq \lambda$ and $\{T_\lambda = \infty\} = \{\langle M \rangle_\infty \leq \lambda\}$, which follows at once from the definition of T_λ. Therefore, we have

$$
\begin{aligned}
E[\mathcal{E}(M)_{T_\lambda} : T_\lambda < \infty] &= E[\mathcal{E}(M)_{T_\lambda}] - E[\mathcal{E}(M)_\infty : T_\lambda = \infty] \\
&= 1 - E[\mathcal{E}(M)_\infty : \langle M \rangle_\infty \leq \lambda],
\end{aligned}
$$

completing the proof. $\qquad\qquad\square$

Example 1.9.(L. A. Shepp [73]) Let $f : R_+ \to R_+$ be a continuous function such that $f(0) > 0$ and $\lim_{t \to \infty} f(t)/t = 0$, and let us define the stopping time

$$
\tau = \inf\{t > 0 : B_t = f(t)\}.
$$

Consider now the martingale $M = B^\tau$ and the stopping time T_λ given in Lemma 1.5. It is easy to see that

$$
\{T_\lambda < \infty\} \subset \{M_{T_\lambda} < f(\lambda), \langle M \rangle_{T_\lambda} = \lambda\}.
$$

Then we have

$$
E[\mathcal{E}(M)_{T_\lambda} : T_\lambda < \infty] \leq \exp\left\{ f(\lambda) - \frac{1}{2}\lambda \right\} \to 0 \quad (\lambda \to \infty).
$$

Thus $\mathcal{E}(M) \in \mathcal{M}_u$ by Lemma 1.5.

For simplicity let us assume that $M \in \mathcal{M}_u$, and let $\varphi : \mathbb{R}_+ \mapsto \mathbb{R}_+$ be a continuous function such that $\varphi(0) = 0$. Then we set

$$
\begin{aligned}
G_\alpha(t, \varphi) &= \exp\left\{ \alpha M_t + \left(\frac{1}{2} - \alpha \right) \langle M \rangle_t - |1 - \alpha|\varphi(\langle M \rangle_t) \right\} \quad (0 \leq t < \infty) \\
g(\alpha, \varphi) &= \sup_T E[G_\alpha(T, \varphi)],
\end{aligned}
$$

where α is a real number and the supremum is taken over all stopping times T.

Lemma 1. 6. *Let $\alpha < \beta < 1$. Then*

(1.11) $$ g(\beta, \varphi) \leq g(\alpha, \varphi)^{(1-\beta)/(1-\alpha)}. $$

On the other hand, if $1 < \alpha < \beta$, we have

(1.12) $$ g(\alpha, \varphi) \leq g(\beta, \varphi)^{(\alpha-1)/(\beta-1)}. $$

Proof. We first show (1.11). For that, let $\alpha < \beta < 1$, and set $p = (1 - \alpha)/(1 - \beta)$, which is larger than 1. Then, the exponent conjugate q to p being $(1 - \alpha)/(\beta - \alpha)$, and

$$
G_\beta(T, \varphi) = G_\alpha(T, \varphi)^{\frac{1}{p}} \mathcal{E}(M)_T^{\frac{1}{q}}.
$$

So, applying Hölder's inequality to the right hand side gives

(1.13) $$ E[G_\beta(T, \varphi)] \leq E[G_\alpha(T, \varphi)]^{\frac{1}{p}} E[\mathcal{E}(M)_T]^{\frac{1}{q}} \leq E[G_\alpha(T, \varphi)]^{\frac{1}{p}}. $$

Consequently, it follows that $g(\beta, \varphi) \leq g(\alpha, \varphi)^{\frac{1}{p}}$.

Secondly, we show (1.12). If $1 < \alpha < \beta$, then

$$G_\alpha(T,\varphi) = G_\beta(T,\varphi)^{(\alpha-1)/(\beta-1)} \mathcal{E}(M)_T^{(\beta-\alpha)/(\beta-1)}$$

and we apply Hölder's inequality with exponents $(\beta-1)/(\alpha-1)$ and $(\beta-1)/(\beta-\alpha)$ to the right hand side :

(1.14) $E[G_\alpha(T,\varphi)] \leq E[G_\beta(T,\varphi)]^{(\alpha-1)/(\beta-1)} E[\mathcal{E}(M)_T]^{(\beta-\alpha)/(\beta-1)}.$

Thus, (1.12) is obtained.   □

First of all, in order to explain our idea explicitly, we shall deal with the special case where $\varphi = 0$: it is rather complicated to deal with the general case. Observe that $\mathcal{E}(M)_t = G_1(t,0)$, $\exp(\frac{1}{2}\langle M\rangle_t) = G_0(t,0)$ and $\exp(\frac{1}{2}M_t) = G_{\frac{1}{2}}(t,0)$. Therefore, Theorem 1.6 means that if $g(\frac{1}{2},0) < \infty$, then $\mathcal{E}(M) \in \mathcal{M}_u$, and its corallary means that $g(\frac{1}{2},0) < \infty$ if $g(0,0) < \infty$.

Lemma 1. 7.   *Let $\alpha \neq 1$. If $g(\alpha,0) < \infty$, then $\mathcal{E}(\alpha M) \in \mathcal{M}_u$.*

Proof.   For $\lambda > 0$ let $T_\lambda = \inf\{t \geq 0 : \langle M\rangle_t > \lambda\}$ as in Lemma 1.5. Since $G_\alpha(t,0) = \mathcal{E}(\alpha M)_t \exp\{\frac{1}{2}(1-\alpha)^2\langle M\rangle_t\}$ and $\langle M\rangle_{T_\lambda} = \lambda$ on $\{T_\lambda < \infty\}$, we find

$$
\begin{aligned}
E[\mathcal{E}(\alpha M)_{T_\lambda}; T_\lambda < \infty] &= E[G_\alpha(T_\lambda,0); T_\lambda < \infty] \exp\left\{-\frac{1}{2}\lambda(1-\alpha)^2\right\} \\
&\leq g(\alpha,0) \exp\left\{-\frac{1}{2}\lambda(1-\alpha)^2\right\},
\end{aligned}
$$

which converges to 0 as $\lambda \to \infty$. Thus $\mathcal{E}(\alpha M)$ is a uniformly integrable martingale by Lemma 1.5.   □

Example 1.10.   For $0 < a < \infty$, we define the stopping time

(1.15) $\tau_a = \inf\{t \geq 0; B_t \leq t - \varphi(t) - a\}.$

Let now $M = B^{\tau_a}$. Then $\frac{1}{2}\langle M\rangle_T - \varphi(\langle M\rangle_T) \leq M_T - \frac{1}{2}\langle M\rangle_T + a$ by the definition of τ_a, so that $g(0,\varphi) \leq e^a \sup_T E[\mathcal{E}(M)_T] \leq e^a$. Furthermore, combining this fact with (1.11) we have $g(\alpha,\varphi) < \infty$ for all α with $0 \leq \alpha < 1$. Therefore, letting $\varphi = 0$, $\mathcal{E}(\alpha M) \in \mathcal{M}_u$ for $0 \leq \alpha < 1$.

Example 1.11.   For $0 < a < \infty$, let ν_a be the stopping time given by

(1.16) $\nu_a = \inf\{t \geq 0; B_t \geq t + \varphi(t) + a\},$

and consider the martingale $M = B^{\nu_a}$. Then it follows from the definition of ν_a that for any stopping time T

$$2M_T + \left(\frac{1}{2} - 2\right)\langle M\rangle_T - \varphi(\langle M\rangle_T) \leq M_T - \frac{1}{2}\langle M\rangle_T + a.$$

Thus $g(2,\varphi) \leq e^a$. Combinig this with (1.12) for all β with $1 < \beta \leq 2$ we have $g(\beta,\varphi) < \infty$, and so, letting $\varphi = 0$, it follows that $\mathcal{E}(\beta M) \in \mathcal{M}_u$ for $1 < \beta \leq 2$.

The next result is an improvement of Lemma 1.7.

Theorem 1. 11.   *If $g(\alpha, 0) < \infty$ for some $\alpha \neq 1$, then $\mathcal{E}(M)$ is a uniformly integrable martingale.*

Proof.   Suppose first that $g(\alpha, 0) < \infty$ for some α with $-\infty < \alpha < 1$. If $\alpha \leq \beta < 1$, then $g(\beta, 0) < \infty$ by (1.11) and so $\mathcal{E}(\beta M) \in \mathcal{M}_u$ by Lemma 1.7. Moreover, since $\mathcal{E}(\beta M)_T \leq G_\beta(T, 0)$ for any stopping time T, it follows from (1.13) that

$$1 = E[\mathcal{E}(\beta M)_\infty] \leq g(\alpha, 0)^{(1-\beta)/(1-\alpha)} E[\mathcal{E}(M)_\infty]^{(\beta-\alpha)/(1-\alpha)}.$$

The last expression converging to $E[\mathcal{E}(M)_\infty]$ as $\beta \uparrow 1$, we find that $1 \leq E[\mathcal{E}(M)_\infty]$. This implies that $\mathcal{E}(M) \in \mathcal{M}_u$.
The proof for the case where $g(\alpha, 0) < \infty$ for some α with $1 < \alpha < \infty$ is completely analogous except that (1.12) is used instead of (1.11).   □

It is necessary to compare the two conditions (a) $g(\alpha, 0) < \infty$ for some $\alpha < 1$ and (b) $g(\beta, 0) < \infty$ for some $\beta > 1$. We shall later show that one of these conditions does not always imply the other.

Recall that a positive continuous function φ is said to be a *lower function* if $P(B_t < \varphi(t), t \to \infty) = 0$; as is well-known, by Blumenthal's zero-one law this probability is equal to 0 or 1.

The main result of this section is the following.

Theorem 1. 12.   *Let φ be a lower function. If $g(\alpha, \varphi) < \infty$ for some $\alpha \neq 1$, then $\mathcal{E}(M)$ is a uniformly integrable martingale.*

Of course, Theorem 1.11 corresponds to the special case $\varphi = 0$. T. Okada proved in [66] that if for some α with $0 \leq \alpha < 1$ the family $\{G_\alpha(T, C\sqrt{t})\}_T$ is uniformly integrable, then $\mathcal{E}(M) \in \mathcal{M}_u$.

On the other hand, by Kolmogorov's criterion, for any positive continuous function φ satisfying $\varphi(t)/t \downarrow$ and $\varphi(t)/\sqrt{t} \uparrow$ as $t \to \infty$, $P(B_t < \varphi(t), t \to \infty) = 0$ or 1 according as

$$\int^{+\infty} t^{-\frac{3}{2}} \varphi(t) \exp\left(-\frac{1}{2t}\varphi(t)^2\right) dt \quad \text{diverges or converges}$$

(see Section 1.8 in [27]). For example, $C\sqrt{t}$ and $\sqrt{2t \log \log t}$ are lower functions. Therefore, as a special case the following corollary contains the result which is an improvement of Okada's criterion.

Corollary 1. 3.   *If $g(\alpha, C\sqrt{t}) < \infty$ for some $\alpha \neq 1$, then $\mathcal{E}(M)$ is a uniformly integrable martingale.*

In order to prove Theorem 1.12, we need four lemmas.

Lemma 1. 8.   *Let τ_a and ν_a be the stopping times by given (1.15) and (1.16) respectively. Then we have*

$$(1.17) \qquad E\left[\exp\left(B_{\tau_a} - \frac{1}{2}\tau_a\right); \tau_a < \infty\right] = P(\tilde{\tau}_a < \infty)$$

$$(1.18) \qquad E\left[\exp\left(B_{\nu_a} - \frac{1}{2}\nu_a\right); \nu_a < \infty\right] = P(\tilde{\nu}_a < \infty)$$

where $\tilde{\tau}_a = \inf\{t \geq 0; B_t \leq -\varphi(t) - a\}$ and $\tilde{\nu}_a = \inf\{t \geq 0; B_t \geq \varphi(t) + a\}$.

Proof.   We show only (1.17), because the proof of (1.18) is similar. Let $N = B^{\tau_a}$. The exponential of N is a martingale, because $\langle N \rangle_t \leq t$ for every t. Then it follows from Girsanov's theorem that for each t the process $\{B_s - s \wedge t \wedge \tau_a\}_{0 \leq s < \infty}$ is a Brownian motion relative to the new probability measure $\mathcal{E}(N)_t dP$. Let now $\Gamma_j : 0 < t_1^{(j)} < t_2^{(j)} < \cdots < t_{n_j}^{(j)} < t$ $(j = 1, 2, \cdots)$ be a sequence of refining partitions which become dense in $[0, t]$. Then we find

$$E[\mathcal{E}(N)_t; t \leq \tau_a]$$
$$= \lim_{j \to \infty} E\left[\mathcal{E}(N)_t; B_{t_k^{(j)}} > t_k^{(j)} - \varphi(t_k^{(j)}) - a, \; 1 \leq k \leq n_j, t \leq \tau_a\right]$$
$$= \lim_{j \to \infty} P\left\{B_{t_k^{(j)}} > -\varphi(t_k^{(j)}) - a, 1 \leq k \leq n_j\right\}$$
$$= P\{B_s > -\varphi(s) - a, 0 \leq \forall s \leq t\},$$

from which $\lim_{t \to \infty} E[\mathcal{E}(N_t); t \leq \tau_a] = P(\tilde{\tau}_a = \infty)$. Therefore, we have

$$P(\tilde{\tau}_a < \infty) = \lim_{t \to \infty} E[\mathcal{E}(N_t); \tau_a < t]$$
$$= E\left[\exp\left(B_{\tau_a} - \frac{1}{2}\tau_a\right); \tau_a < \infty\right],$$

completing the proof.   □

Lemma 1. 9.   *Let $0 < a < \infty$. If φ is a lower function, so is $\varphi + a$.*

Proof.   Consider the stopping times

$$\mu = \inf\left\{t > 0; B_t \geq t\varphi\left(\frac{1}{t}\right) + a(1 \wedge t)\right\},$$

$$\sigma = \inf\left\{t > 0; B_t \geq t\varphi\left(\frac{1}{t}\right) + at\right\}.$$

It is clear that $\{\mu = 0\} = \{\sigma = 0\}$. Observe that $\{tB_{1/t}\}$ is also a Brownian motion. Then

$$P\{B_t < \varphi(t) + a, t \to \infty\} = P\{B_t < t\varphi(1/t) + at, t \to 0\}$$
$$= P(\sigma > 0)$$
$$= P(\mu > 0).$$

On the other hand, by the theorem of Girsanov the process $\hat{B}$ defined by the formula

$$\hat{B}_t = B_t - a(1 \wedge t) \quad (0 \leq t < \infty)$$

is a Brownian motion under $\hat{P}$ where $d\hat{P} = \exp(aB_1 - \frac{1}{2}a^2)dP$. Now, let us assume that φ is a lower function. Then

$$\hat{P}(\mu > 0) = \hat{P}\left\{\hat{B}_t < t\varphi\left(\frac{1}{t}\right), t \to 0\right\} = 0$$

and, since $\hat{P}$ is equivalent to P, we have $P(\mu > 0) = 0$. This completes the proof. □

Lemma 1. 10.   *φ is a lower function if and only if $P(\tilde{\tau}_a < \infty) = 1$ for all $a > 0$.*

Proof. Suppose first that φ is a lower function. Then $\varphi + a$ is also a lower function by Lemma 1.9. So, we have

$$P(\tilde{\tau}_a = \infty) = P\{B_t < \varphi(t) + a, 0 \leq \forall t < \infty\} = 0.$$

Conversely, let us assume that $P(\tilde{\tau}_a < \infty) = 1$ for all $a > 0$. Then, since $\{tB_{1/t}\}$ is a Brownian motion, we find

$$(1.19) \qquad P\{B_t < t\varphi(1/t) + at, 0 < \forall t < \infty\} = P(\tilde{\tau}_a = \infty) = 0.$$

Suppose now that φ is not a lower function. Then by Blumenthal's zero-one law $P\{B_t < t\varphi(1/t), t \to 0\} = 1$. In other words, $P(\tau > 0) = 1$ where $\tau = \inf\{t > 0 : B_t \geq t\varphi(1/t)\}$. Let u be a positive number such that $P(\tau > u) > \frac{1}{2}$; namely, $P\{B_s < s\varphi(1/s), 0 < \forall s \leq u\} > \frac{1}{2}$. On the other hand, $P(B_s < a, a \to 0) = 1$ and so $P(\sigma_a > 0) = 1$ where $\sigma_a = \inf\{t \leq 0 : B_t \geq a\}$. Furthermore, $P(\sigma_a > 1/u) > \frac{1}{2}$ for large a, because $\sigma_a \to \infty$ as $a \to \infty$ from the definition of σ_a. This means

$$P(B_s < as, u \leq \forall s < \infty) = P\left(B_t < a, 0 < \forall t \leq \frac{1}{u}\right) = P\left(\sigma_a > \frac{1}{u}\right) > \frac{1}{2}.$$

Then, combining these facts we find

$$P(B_t < t\varphi(1/t) + at, 0 < \forall t < \infty)$$
$$\geq P(B_t < t\varphi(1/t) \text{ for } \forall t < u \text{ and } B_t < at \text{ for } \forall t \geq u) > 0,$$

which is inconsistent with (1.19). Consequently, φ is a lower function. $\qquad\square$

By using the same argument we obtain the following.

Lemma 1. 11. *φ is a lower function if and only if $P(\tilde{\nu}_a < \infty) = 1$ for all $a > 0$.*

Proof of Theorem 1.12: As is well-known, any continuous local martingale can be reduced to stopped Brownian motion by means of a continuous change of time. Therefore, it suffices essentially to verify this theorem in the case where $M = B^\zeta$ for a certain stopping time ζ.

We begin with the case where $g(\alpha, \varphi) < \infty$ for some α with $-\infty < \alpha < 1$. Let

$$\tau_j = \inf\{t \geq 0; B_t \leq t - \varphi(t) - j\} \quad (j = 1, 2, \cdots),$$

which is nothing but the stopping time obtained by letting $a = j$ in (1.15). Then, since φ is a lower function by the assumption, combining Lemmas 1.8 and 1.10 we have

$$E\left[\exp\left(B_{\tau_j} - \frac{1}{2}\tau_j\right)\right] = 1$$

and so

$$1 = E\left[\exp\left(B_{\zeta \wedge \tau_j} - \frac{1}{2}(\zeta \wedge \tau_j)\right)\right] \leq E[\mathcal{E}(B^\zeta)_\infty] + E\left[\exp\left(B_{\tau_j} - \frac{1}{2}\tau_j\right); \tau_j < \zeta\right].$$

Noticing that $B_{\tau_j} = \tau_j - \varphi(\tau_j) - j$ on $\{\tau_j < \infty\}$, we find that the expectation in the last expression is smaller than

$$E[G_\alpha(\tau_j, \varphi) \exp\{(1 - \alpha)(B_{\tau_j} - \tau_j - \varphi(\tau_j))\}; \tau_j < \infty]$$
$$\leq g(\alpha, \varphi) \exp(-(1 - \alpha)j),$$

which tends to 0 as $j \to \infty$, because $g(\alpha, \varphi) < \infty$ by the assumption. Thus $1 \leq E[\mathcal{E}(B^\zeta)_\infty]$, from which $\mathcal{E}(B^\zeta) \in \mathcal{M}_u$.

Secondly, we deal with the case where $g(\beta, \varphi) < \infty$ for some β with $1 < \beta < \infty$. Instead of τ_j, consider this time the stopping time

$$\nu_j = \inf\{t \geq 0; B_t \geq t + \varphi(t) + j\}.$$

Then $P(\tilde{\nu}_j < \infty) = 1$ $(j \geq 1)$ by Lemma 1.11, because φ is a lower function. Therefore, combining this fact with Lemma 1.8 we have

$$1 \leq E[\mathcal{E}(B^\zeta)_\infty] + E\left[\exp\left(B_{\nu_j} - \frac{1}{2}\nu_j\right); \nu_j < \zeta\right].$$

Since $B_{\nu_j} = \nu_j + \varphi(\nu_j) + j$ on $\{\nu_j < \infty\}$, the second term on the right hand side is smaller than

$$E[G_\beta(\nu_j, \varphi)\exp\{-(\beta-1)(B_{\nu_j} - \nu_j - \varphi(\nu_j))\}; \nu_j < \infty]$$
$$\leq g(\beta, \varphi)\exp\{-(\beta-1)j\} \to 0 \quad (j \to \infty).$$

Consequently, it follows that $1 \leq E[\mathcal{E}(B^\zeta)]$. This completes the proof. □

Remark 1.4.   Let φ be a lower function as above. If $g(\alpha, \varphi) < \infty$ for some α with $-\infty < \alpha < 1$ (resp. $1 < \alpha < \infty$),then the family $\{G_\beta(T, \varphi)\}_T$ is uniformly integrable for every β with $\alpha < \beta < 1$ (resp. $1 < \beta < \alpha$). In fact, let $\alpha < \beta < 1$. Then for any $\lambda > 0$ and any stopping time T we can obtain

$$E[G_\beta(T, \varphi) : G_\beta(T, \varphi) > \lambda]$$
$$\leq g(\alpha, \varphi)^{(1-\beta)/(1-\alpha)} E[\mathcal{E}(M)_T : G_\beta(T, \varphi) > \lambda]^{(\beta-\alpha)/(1-\alpha)}$$

by modifying slightly the proof of (1.13). Assume now that $g(\alpha, \varphi) < \infty$. Then $\mathcal{E}(M) \in \mathcal{M}_u$ by Theorem 1.12, so that

$$E[\mathcal{E}(M)_T; G_\beta(T, \varphi) > \lambda] = E[\mathcal{E}(M)_\infty; G_\beta(T, \varphi) > \lambda].$$

Furthermore, by using Chebyshev's inequality and then the inequality (1.11) we find

$$\lambda P\{G_\beta(T, \varphi) > \lambda\} \leq g(\alpha, \varphi)^{(1-\beta)/(1-\alpha)} \quad (\lambda > 0).$$

From these estimations the uniform integrability of the family $\{G_\beta(T, \varphi)\}_T$ follows immediately.

On the other hand, in order to give the proof for the case where $g(\alpha, \varphi) < \infty$ for some α with $1 < \alpha < \infty$ it suffices to apply (1.12) and (1.14) instead of (1.11) and (1.13).

Now, let us consider the class Φ of positive continuous functions φ satisfying

$$\liminf_{t \to \infty} \varphi(t)/t = 0.$$

Obviously, it is larger than the class of all lower functions. But the reverse inclusion fails. For example, $\varphi(t) = (1 + \varepsilon)\sqrt{2t \log\log t}(\varepsilon > 0)$ belongs to the class Φ, but it is not a lower function by Kolmogorov's criterion.

Theorem 1.13. *Let $\varphi \in \Phi$. If $g(\alpha, \varphi) < \infty$ for $\alpha \neq 1$, then $\mathcal{E}(\alpha M) \in \mathcal{M}_u$.*

Proof. For $\lambda > 0$, let $T_\lambda = \inf\{t \geq 0 : \langle M \rangle_t > \lambda\}$ as before. Since $G_\alpha(t, \varphi) = \mathcal{E}(\alpha M)_t \exp\{(1 - \alpha)^2 \langle M \rangle_t / 2 - |1 - \alpha|\varphi(\langle M \rangle_t)\}$, we find

$$E[\mathcal{E}(\alpha M)_{T_\lambda} : T_\lambda < \infty]$$
$$= E[G_\alpha(T_\lambda, \varphi) : T_\lambda < \infty] \exp\left\{|1 - \alpha|\varphi(\lambda) - \frac{1}{2}\lambda(1 - \alpha)^2\right\}$$
$$\leq g(\alpha, \varphi) \exp\left\{-\lambda|1 - \alpha|\left(\frac{1}{2}|1 - \alpha| - \frac{\varphi(\lambda)}{\lambda}\right)\right\}.$$

Moreover, $\varphi(\lambda)/\lambda$ converges to 0 as $\lambda \to \infty$ by the assumption. Therefore, we have $\liminf_{\lambda \to \infty} E[\mathcal{E}(\alpha M)_{T_\lambda} : T_\lambda < \infty] = 0$. Then $\mathcal{E}(\alpha M) \in \mathcal{M}_u$ by Lemma 1.5. Thus the proof is complete. $\square$

As an illustration, consider the case where $g(\alpha, \varphi) < \infty$ for some α with $-\infty < \alpha < 1$. If $\alpha \leq \beta < 1$, then $g(\beta, \varphi) < \infty$ by (1.11) and so $\mathcal{E}(\beta M) \in \mathcal{M}_u$ by Theorem 1.13. However, there are some cases where $\mathcal{E}(M) \notin \mathcal{M}_u$ in addition to that. For example, for $\varepsilon > 0$ let $\varphi(t) = (1 + \varepsilon)\sqrt{2t \log \log t}$ and let τ_a be the corresponding stopping time defined by (1.15). Consider now $M = B^{\tau_a}$. Then $g(0, \varphi) < \infty$ by Example 1.10, and so $\mathcal{E}(\beta M) \in \mathcal{M}_u$ for all β with $0 \leq \beta < 1$ by Theorem 1.13. On the other hand, since φ is not a lower function, it follows from Lemmas 1.8 and 1.10 that $E[\mathcal{E}(M)_\infty] < 1$. Namely, $\mathcal{E}(M) \notin \mathcal{M}_u$.

By contrast, considering the martingale $M = B^{\nu_a}$ where ν_a denotes the stopping time defined by (1.16), we can obtain an example such that $\mathcal{E}(\beta M) \in \mathcal{M}_u$ for all β with $1 < \beta \leq 2$ and $\mathcal{E}(M) \notin \mathcal{M}_u$.

We close this section with three further counterexamples.

Example 1.12. Let $\tau = \inf\{t \geq 0 : B_t \leq t - 1\}$, which is nothing but the stopping time defined by setting $\varphi = 0$ and $a = 1$ in (1.15). Consider now the martingale $M = B^\tau$. Then $g(0, 0) \leq e$ by Example 1.10, so that $\mathcal{E}(\beta M) \in \mathcal{M}_u$ for all β with $0 \leq \beta \leq 1$ according to Theorem 1.12 and Theorem 1.13. On the other hand, since $\mathcal{E}(aM)_\infty = \exp(aB_\tau - a^2\tau/2) \leq e^{-a}\exp(\tau/2)$ on $\{\tau < \infty\}$, we find that for $a > 1$

$$E[\mathcal{E}(\alpha M)_\infty] \leq e^{-a}g(0, 0) \leq e^{1-a} < 1.$$

Namely, $\mathcal{E}(aM) \notin \mathcal{M}_u$ for $a > 1$. This implies that $g(\beta, \psi) = \infty$ for any $\psi \in \Phi$ and any number $\beta > 1$.

Example 1.13. Let now $M = B^\nu$, where $\nu = \inf\{t \geq 0 : B_t \geq t + 1\}$. Obviously, ν is the stopping time obtained by setting $\varphi = 0$ and $a = 1$ in (1.16). Then $g(2, 0) < \infty$ by Example 1.11. Therefore, from Theorem 1.12 and Theorem 1.13 it follows that $\mathcal{E}(\beta M) \in \mathcal{M}_u$ for all β with $1 \leq \beta \leq 2$. On the other hand, since $\mathcal{E}(M)_\infty = \exp(1 + \nu/2)$ on $\{\nu < \infty\}$, we find

$$E[\mathcal{E}(aM)_\infty : \nu < \infty] = E\left[\exp\left\{\frac{1}{2}a(2 - a)\nu + a\right\} : \nu < \infty\right]$$
$$\leq e^a E\left[\exp\left(\frac{1}{2}\nu\right) : \nu < \infty\right] \leq e^{a-1}.$$

Thus $\mathcal{E}(aM) \notin \mathcal{M}_u$ if $a < 1$. This implies that $g(\alpha, \psi) = \infty$ for any $\psi \in \Phi$ and any number $\alpha < 1$. By this example it comes out that Theorem 1.12 properly contains the criteria of Okada and Novikov.

Example 1.14. Let $B = (B_t)$ and $\tilde{B} = (\tilde{B}_t)$ be two independent Brownian motions, and define the stopping times $\tau = \inf\{t \geq 0 : B_t \leq t - 1\}$ and $\nu = \inf\{t \geq 0 : \tilde{B}_t \geq t+1\}$ as before. We have already seen in Examples 1.12 and 1.13 that the exponential processes $\mathcal{E}(B^\tau)$ and $\mathcal{E}(\tilde{B}^\nu)$ are uniformly integrable martingales. Consider now the continuous martingale $M = B^\tau + \tilde{B}^\nu$. From the independence of B and $\tilde{B}$ it follows immediately that $\langle M \rangle_t = t \wedge \tau + t \wedge \nu$. Then $\mathcal{E}(M) = \mathcal{E}(B^\tau)\mathcal{E}(\tilde{B}^\nu)$, and so by the independence we have

$$E[\mathcal{E}(M)] = E[\mathcal{E}(B^\tau)]E[\mathcal{E}(\tilde{B}^\nu)] = 1.$$

Namely, $\mathcal{E}(M) \in \mathcal{M}_u$. However, Example 1.12 implies that $E[\mathcal{E}(aB^\tau)_\infty] < 1$ for $a > 1$. On the other hand, $E[\mathcal{E}(a\tilde{B}^\nu)_\infty] < 1$ for $a < 1$ by Example 1.13. From these facts it follows that $\mathcal{E}(aM) \notin \mathcal{M}_u$ for $a \neq 1$, which implies that $g(\alpha, \varphi) = \infty$ for any number $\alpha \neq 1$ and any function $\varphi \in \Phi$. Thus the converse of Theorem 1.12 does not hold.

Chapter 2

BMO-Martingales

2.1 The class BMO

A real-valued locally integrable function f defined on $\mathbb{R}^n$ is said to be in BMO, the space of functions of bounded mean oscillation, if

$$\sup_Q \frac{1}{|Q|} \int_Q |f(x) - f_Q| dx < \infty$$

where the supremum is taken over all cubes Q in $\mathbb{R}^n$, $|Q|$ denotes the volume of Q, and

$$f_Q = \frac{1}{|Q|} \int_Q f(x) dx.$$

The space of BMO-functions was introduced by F. John and L. Nirenberg ([31]) and they gave the first important result on BMO-functuins. In 1971 C. Fefferman ([17]) characterize the space of BMO- functions as the dual of the Hardy space H_1. This is one of the most important results in the theory of H_1. On the other hand, in 1972 R. K. Getoor and M. J. Sharpe ([21]) introduced the concept of a conformal martingale and by using conformal martingales they established the duality of H_1 and BMO in the probabilistic setting.

Let $M = (M_t, \mathcal{F}_t)$ be a uniformly integrable martingale with $M_0 = 0$, and for $1 \le p < \infty$ we set

$$(2.1) \qquad \| M \|_{BMO_p} = \sup_T \left\| E[|M_\infty - M_{T-}|^p|\mathcal{F}_T]^{1/p} \right\|_\infty$$

where the supremum is taken over all stopping times T. The class $\{M : \|M\|_{BMO_p} < \infty\}$ is denoted by BMO_p, and observe that $\| \quad \|_{BMO_p}$ is a norm on this space. From the Hölder inequality it follows at once that for $p < q$, $BMO_q \subset BMO_p$. The reverse inclusion will be proved in the next section, so we write simply BMO for this space. Note that, if $M \in BMO$ and T is a stopping time, $M^T \in BMO$ and $\|M^T\|_{BMO_1} \le \|M\|_{BMO_1}$. It will follow from Theorem 2.6 that the space BMO, which can be identified with the dual space of H_1, is complete.

As stated before, we deal entirely with continuous local martingales. Observe that if $\|M\|_2 < \infty$, then

$$\| M \|_{BMO_2} = \sup_T \left\| E[\langle M \rangle_\infty - \langle M \rangle_T|\mathcal{F}_T]^{1/2} \right\|_\infty,$$

which follows from the fact that $M^2 - \langle M \rangle$ is a uniformly integrable martingale. On the other hand, the energy inequalities give

$$E[\langle M \rangle_\infty^n] \leq n! \|M\|_{BMO_2}^{2n} \quad (n = 1, 2, \cdots),$$

so that $BMO \subset H_p$ for every p. Now, let L_∞ be the class of all bounded martingales and let H_∞ be the class of all martingales M such that $\langle M \rangle_\infty$ is bounded. Since $\|M\|_{BMO_1} \leq 2\|M\|_\infty$ and $\|M\|_{BMO_2} \leq \|\langle M \rangle_\infty\|_\infty^{1/2}$, these two classes L_∞ and H_∞ are obviously contained in BMO. However, as is easily seen, BMO is neither L_∞ nor H_∞ in general, and further there is not an inclusion relation between L_∞ and H_∞. For example, if $B = (B_t)$ is a one dimensional Brownian motion, the process B stopped at τ, where $\tau = \inf\{t : |B_t| = 1\}$, belongs to $L_\infty \setminus H_\infty$. On the other hand, one can easily see that $(B_{t \wedge 1}) \in H_\infty \setminus L_\infty$.

Observe now that the martingale $E[B_1^2|\mathcal{F}_t]$ is not in BMO. Incidentally, if $M, N \in BMO$ and if f is a Lipschitz function on $\mathbb{R}^2$, that is, $|f(x,y) - f(x',y')| \leq c(|x - x'| + |y - y'|)$ for all $(x,y),(x',y') \in \mathbb{R}^2$, with a constant $c > 0$, then the martingale L defined by $L_t = E[f(M_\infty, N_\infty)|\mathcal{F}_t]$ $(0 \leq t < \infty)$ is also in BMO. In fact, for any stopping time T we have

$$
\begin{aligned}
E[|L_\infty - L_T|\,|\mathcal{F}_T] &\leq 2E[|f(M_\infty, N_\infty) - f(M_T, N_T)|\,|\mathcal{F}_T] \\
&\leq 2c(E[|M_\infty - M_T|\,|\mathcal{F}_T] + E[|N_\infty - N_T|\,|\mathcal{F}_T]) \\
&\leq 2c(\|M\|_{BMO_1} + \|N\|_{BMO_1}).
\end{aligned}
$$

Particulary, the martingales $E[M_\infty \vee N_\infty|\mathcal{F}.]$ and $E[M_\infty \wedge N_\infty|\mathcal{F}.]$ belong to the class BMO. Furthermore, it follows that for each $p \geq 1$

$$(2.2) \qquad\qquad \|M\|_{BMO_p} = \sup_T \frac{\|M_\infty - M_T\|_p}{P(T < \infty)^{1/p}}.$$

where the supremum is taken over all stopping times T satisfying $P(T < \infty) > 0$. To show this, let α_p be the right-hand side of (2.2) for simplicity. Then by the definition of $\|\ \|_{BMO_p}$

$$
\begin{aligned}
E[|M_\infty - M_T|^p] &= E[E[|M_\infty - M_T|^p|\mathcal{F}_T]\mathrm{I}_{\{T<\infty\}}] \\
&\leq \|M\|_{BMO_p}^p P(T < \infty),
\end{aligned}
$$

which implies that $\alpha_p \leq \|M\|_{BMO_p}$. On the other hand, for each $A \in \mathcal{F}_T$ we have

$$
\begin{aligned}
E[E[|M_\infty - M_T|^p|\mathcal{F}_T] : A] &= E[|M_\infty - M_{T_A}|^p] \\
&\leq \alpha_p^p P(T_A < \infty) \\
&\leq \alpha_p^p P(A),
\end{aligned}
$$

from which it follows at once that $\|M\|_{BMO_p} \leq \alpha_p$.

We close this section with a nice remark given by M. Emery. One can easily verify that, if $M \in BMO$, then for every ε with $0 < \varepsilon < 1$ there is a constant $a > 0$ satisfying the inequality

$$(2.3) \qquad\qquad P\left(\sup_{0 \leq t < \infty} |M_{T+t} - M_T| > a|\mathcal{F}_T\right) \leq 1 - \varepsilon$$

for any stopping time T. M. Emery showed in [15] that, if there exist two constants $a \geq 0$ and $0 < \varepsilon < 1$ such that the inequality (2.3) is valid for every stopping time T, then M belongs to the class BMO.

2.2 The John-Nirenberg inequality

The next inequality, which is called the John-Nirenberg inequality, is of fundamental importance in our investigation.

Theorem 2. 1. *If* $\| M \|_{BMO_1} < 1/4$, *then for any stopping time* T

$$(2.4) \qquad E[\exp(|M_\infty - M_T|)|\mathcal{F}_T] \leq \frac{1}{1 - 4\| M \|_{BMO_1}}.$$

Note that the sample continuity of M is not essential for the validity of this inequality. To prove this theorem, we use the next lemma.

Lemma 2. 1 (D. W. Stroock [79]). *Let* $X = (X_t, \mathcal{F}_t)_{0 \leq t \leq \infty}$ *be a right continuous process with left hand limits. If there is a non- negative integrable random variable* U *such that*

$$(2.5) \qquad E[\|X_T - X_{S-}\||\mathcal{F}_S] \leq E[U|\mathcal{F}_S]$$

for any pair S, T *of stopping times with* $S \leq T$, *then the inequality*

$$(2.6) \qquad \beta P(X^* \geq \alpha + \beta) \leq E[U : X^* \geq \alpha]$$

holds for every $\alpha, \beta > 0$. *Here* $X^* = \sup_{0 \leq t \leq \infty} |X_t|$.

Proof. Let $\alpha, \beta > 0$ be given, and set

$$S = \inf\{t \geq 0 : |X_t| \geq \alpha\}, \quad T = \inf\{t \geq 0 : |X_t| \geq \alpha + \beta\}.$$

Clearly S and T are stopping times with $S \leq T$ a.s, and from the definitions of these stopping times it follows that

$$\{X^* > \alpha + \beta\} \subset \{|X_T - X_{S-}| \geq \beta, |X_S| \geq \alpha\}$$

and $\{|X_S| \geq \alpha\} \in \mathcal{F}_S$. Then by the assumption we have

$$P(X^* > \alpha + \beta) \leq \frac{1}{\beta} E[\|X_T - X_{S-}\| : |X_S| \geq \alpha]$$
$$\leq \frac{1}{\beta} E[U : X^* \geq \alpha],$$

from which (2.6) follows at once. □

Proof of Theorem 2.1 : Assume that $\| M \|_{BMO_1} < 1/4$, and we set $c = \| M \|_{BMO_1}$ for convenience. Then for any pair S, T of stopping times with $S \leq T$

$$E[\|M_T - M_S\||\mathcal{F}_S] \leq E[\|M_\infty - M_S\||\mathcal{F}_S] \leq c,$$

and so by Lemma 2.1 we have

$$\beta P\{(M^* - \beta)^+ \geq \alpha\} \leq \beta P(M^* \geq \alpha + \beta) \leq cP(M^* \geq \alpha)$$

for every $\alpha, \beta > 0$. From this it follows immediately that $\beta E[(M^* - \beta)^+] \leq cE[M^*]$. Setting $\beta = 2c$, we have $2cE[M^* - 2c] \leq cE[M^*]$, that is, $E[M^*] \leq 4c$. It is not difficult to verify that the conditional form of this inequality is the following : for any stopping time S

$$E\left[\sup_{S \le t < \infty} |M_t - M_S||\mathcal{F}_S\right] \le 4c.$$

Then $E[M^* - M_S^*|\mathcal{F}_S] \le 4c$, so that

$$E\left[\sup_{S \le t < \infty} |M_t - M_S|^n|\mathcal{F}_S\right] \le n!(4c)^n \quad (n = 1, 2, \cdots)$$

by the energy inequalities. Theorefore, the inequality

$$E[\exp(|M_\infty - M_S|)|\mathcal{F}_S] \le \frac{1}{1 - 4c}$$

is valid for any stopping time S. □

Corollary 2. 1.　　*Let $1 < p < \infty$. There is a positive constant C_p such that for any uniformly integrable martingale M*

$$\| M \|_{BMO_1} \le \| M \|_{BMO_p} \le C_p \| M \|_{BMO_1}.$$

Proof.　　The left hand side inequality follows at once from the Jensen inequality. To show the right hand side inequality, it is enough to consider the case where $\| M \|_{BMO_1} > 0$. By Theorem 2.1 we have

$$E\left[\exp\left(\frac{1}{8\| M \|_{BMO_1}}|M_\infty - M_T|\right)\Big|\mathcal{F}_T\right] \le 2,$$

from which the conclusion can be easily drawn. For example, if $p \in \mathbb{N}$, then

$$\frac{1}{p!(8\| M \|_{BMO_1})^p}E[|M_\infty - M_T|^p|\mathcal{F}_T] \le 2,$$

namely, $\| M \|_{BMO_p} \le 8 \cdot 2^{1/p}(p!)^{1/p}\| M \|_{BMO_1}$. □

We state again the definition of a BMO-martingale for form's sake.

Definition 2. 1.　　*Let BMO be the class of all uniformly integrable martingales M such that $\| M \|_{BMO_1} < \infty$. The martingale M in BMO is called a BMO-martingale.*

It is a matter of course that the class BMO depends on the underlying probability measure, and so we shall denote it by $BMO(P)$ in case of necessity.

Remark 2.1.　　Let now Φ be an increasing convex function on $[0, +\infty[$ with $\Phi(0) = 0$, and let $M = (M_t)_{t \ge 0}$ be a uniformly integrable right continuous martingale. We set

$$\|M\|_{BMO_\Phi} = \inf\left\{\lambda > 0 : \sup_T \left\| E\left[\Phi\left(\frac{1}{\lambda}|M_\infty - M_{T-}|\right)\Big|\mathcal{F}_T\right]\right\|_\infty \le 1\right\},$$

where the supremum is taken over all stopping times T. The BMO_Φ norm was introduced in [2] by N. L. Bassily and J. Mogyorodi, and they proved implicitely that if $\int_0^\infty \Phi(ct)e^{-t}dt < +\infty$ for some constant $c > 0$, then for any uniformly integrable right continuous martingale $M = (M_t)$

$$(2.7) \qquad c_\Phi \|M\|_{BMO_1} \leq \|M\|_{BMO_\Phi} \leq C_\Phi \|M\|_{BMO_1},$$

where $c_\Phi > 0$ and $C_\Phi > 0$ depend only on Φ. Quite recently, M. Kikuchi has shown in [50] that if $A = (A_t)$ is an adapted right continuous process whose left potential is bounded by a constant $c > 0$, then the inequality

$$(2.8) \qquad E[\Phi(A_\infty)] \leq \frac{1}{c} \left(\int_0^\infty \Phi(ct)e^{-t}dt \right) E[A_\infty]$$

holds, and by using this inequality he has given a nice proof of the right-hand inequality of (2.7). Following his idea, we shall sketch its proof. Firstly, let

$$C_\Phi = \left[\inf \left\{ c > 0 : \int_0^\infty \Phi(ct)e^{-t}dt > 1 \right\} \right]^{-1}.$$

Then $0 < C_\Phi < \infty$ by the assumption. Since $E[M^* - M_{T-}^*|\mathcal{F}_T] \leq 4\|M\|_{BMO_1}$ for every stopping time T, it follows from (2.8) that

$$E \left[\Phi \left(\frac{1}{4C_\Phi \|M\|_{BMO_1}} M^* \right) \right] \leq \frac{1}{4\|M\|_{BMO_1}} \left(\int_0^\infty \Phi \left(\frac{t}{C_\Phi} \right) e^{-t}dt \right) E[M^*] \leq 1.$$

Putting this in conditional form gives

$$E \left[\Phi \left(\frac{1}{4C_\Phi \|M\|_{BMO_1}} \sup_t |M_{T+t} - M_{T-}| \right) \bigg| \mathcal{F}_T \right] \leq 1.$$

Then $\|M\|_{BMO_\Phi} \leq 4C_\Phi \|M\|_{BMO_1}$. The left-hand inequality is an immediate consequence of Jensen's inequality.

The next inequality, which is also called the John-Nirenberg inequality, was given in [19] by A. M. Garsia for discrete parameter martingales and by P. A. Meyer ([60]) for general martingales.

Theorem 2. 2. *If $\| M \|_{BMO_2} < 1$, then for every stopping time T*

$$(2.9) \qquad E[\exp(\langle M \rangle_\infty - \langle M \rangle_T)|\mathcal{F}_T] \leq \frac{1}{1 - \| M \|_{BMO_2}^2}.$$

Proof. By the definition of $\| M \|_{BMO_2}$ we have

$$E[\langle M \rangle_\infty - \langle M \rangle_T|\mathcal{F}_T] \leq \| M \|_{BMO_2}^2,$$

which follows from the fact that $M^2 - \langle M \rangle$ is a uniformly integrable martingale. Then the energy inequalities give

$$E[(\langle M \rangle_\infty - \langle M \rangle_T)^n|\mathcal{F}_T] \leq n! \| M \|_{BMO_2}^{2n}.$$

Therefore, as $\| M \|_{BMO_2} < 1$, we get

$$
\begin{aligned}
E[\exp(\langle M \rangle_\infty - \langle M \rangle_T)|\mathcal{F}_T] &\leq \sum_{n=0}^{\infty} \frac{1}{n!} E[(\langle M \rangle_\infty - \langle M \rangle_T)^n |\mathcal{F}_T] \\
&\leq \sum_{n=0}^{\infty} \| M \|_{BMO_2}^{2n} \\
&\leq \frac{1}{1 - \| M \|_{BMO_2}^2},
\end{aligned}
$$

completing the proof.   □

The following example shows that this estimate is the best possible.

Example 2.1.   Firstly, let $\mathcal{G}^0$ be the class of all topological Borel sets in $\mathbb{R}_+ = [0, \infty[$, and let S be the identity mapping of $\mathbb{R}_+$ onto $\mathbb{R}_+$. We define a probability measure $d\mu$ on $\mathbb{R}_+$ such that $\mu(S > t) = e^{-t}$. Let $\mathcal{G}$ be the completion of $\mathcal{G}^0$ with respect to $d\mu$, and similarly let $\mathcal{G}_t$ be the completion of the Borel field generated by $S \wedge t$. It is clear that S is a stopping time over $(\mathcal{G}_t)$.
We now construct in the usual way a probability system $(\Omega, \mathcal{F}, P; (\mathcal{F}_t))$ by taking the product of the system $(\mathbb{R}_+, \mathcal{G}, d\mu; (\mathcal{G}_t))$ with another system $(\Omega', \mathcal{F}', P'; (\mathcal{F}'_t))$ which carries a one dimensional Brownian motion $B = (B_t)$ starting at 0. Then S is also a stopping time over $(\mathcal{F}_t)$. Let now $M = B^S/\sqrt{2}$. It is clearly a continuous martingale such that $\langle M \rangle_t = (S \wedge t)/2$. Since $\langle M \rangle_\infty - \langle M \rangle_t = 2^{-1}(S - t)I_{\{t<S\}}$, it is not difficult to see that $\| M \|_{BMO_2}^2 = 1/2$, and so the right hand side of (2.9) is equal to 2. On the other hand, we find that

$$
\begin{aligned}
E[\exp(\langle M \rangle_\infty)] &= E\left[\exp\left(\frac{1}{2}S\right)\right] \\
&= \int_0^\infty \exp\left(\frac{1}{2}x\right)\exp(-x)dx \\
&= 2.
\end{aligned}
$$

Thus the inequality (2.9) can not be improved.

Remark 2.2.   We shall make mention of a relation between a BMO-function and a BMO-martingale. Let $D = \{z : |z| < 1\}$ be the unit disc in the complex plane, ∂D its boundary and $m(d\theta)$ the normalized Lebesgue measure on ∂D. An integrable real valued function f is in BMO if there is a positive constant C such that for all intervals $I \subset \partial D$,

$$
\frac{1}{m(I)} \int_I |f - f_I| m(d\theta) \leq C,
$$

where $f_I = \frac{1}{m(I)} \int_I f\,dm$ is the average value of f on I. The smallest constant with this property is denoted by $\|f\|_{BMO}$. On the other hand, let

$$
h(z) = \int_0^{2\pi} f(t)P(r, \theta - t)m(dt) \quad (z = re^{i\theta} \in D),
$$

where $P(r, \eta) = \dfrac{1 - r^2}{1 - 2r\cos(\eta) + r^2}$ is the *Poisson kernel*. Then h is the harmonic function in D with boundary function f. Let now $B = (B_t, \mathcal{F}_t)$ be the complex Brownian motion starting at 0 and let $\tau = \inf\{t : |B_t| = 1\}$. As is well known, the process $(h(B_{t\wedge\tau}), \mathcal{F}_{t\wedge\tau})$ is a uniformly integrable martingale. In particular, if f is in BMO, then the process $h(B^\tau)$ is a BMO-martingale and there are constants $c, C > 0$, independent of f, such that

$$c\|f\|_{BMO} \leq \|h(B^\tau)\|_{BMO_1} \leq C\|f\|_{BMO}.$$

Conversely, if X is a uniformly integrable martingale adapted to the filtration $(\mathcal{F}_{t\wedge\tau})$, then there is a (unique) Borel measurable function f defined on ∂D such that $f(B_\tau) = E[X_\infty|\sigma(B_\tau)]$. Let us consider the mapping $J : X \mapsto f$. Then there is a constant C such that

$$\|J(X)\|_{BMO} \leq C\|X\|_{BMO}$$

for all BMO-martingales X adapted to $(\mathcal{F}_{t\wedge\tau})$. The family of all real-valued BMO-functions on ∂D is identified in this way with the family of all BMO-martingales X which have X_∞ measurable with respect to $\sigma(B_\tau)$.

2.3 Characterizations of a BMO-martingale

R. K. Getoor and M. J. Sharpe proved in [21] that the space of BMO-martingales can be characterized as the dual of H_1. It is needless to say that their result is nothing but a probabilistic analogue of C. Fefferman's duality theorem. Furthermore, it was shown in the general setting by C. Herz and independently by D. Lépingle that $M \in BMO$ if and only if there exists a (not necessarily adapted) process $A = (A_t)_{0 \leq t \leq \infty}$ with integrable variation such that

(a) $\int_{[0,\infty]} |dA_t| \leq C$

(b) $M_t = E[A_\infty^o|\mathcal{F}_t]$ $(0 \leq t \leq \infty)$

where A^o denotes the dual optional projection of A. For the proof, see Theorem 78, p.288, in [6].

The aim of this section is to give other characterizations of a BMO-martingale. We begin with a simple remark. Let $S, B = (B_t, \mathcal{F}_t)$ and $(\Omega, \mathcal{F}, P)$ be as in Example 2.1. Then the martingale $M = 2\sqrt{2}B^S$ belongs to the class BMO. But it does not satisfy the condition given in Theorem 1.6, because

$$
\begin{aligned}
E\left[\exp\left(\frac{1}{2}M_\infty\right)\right] &= E[\exp(\sqrt{2}B_S)] \\
&= \int_0^\infty e^t d\mu(t) \int_{\Omega'} \exp(\sqrt{2}B_t - t)dP' \\
&= \int_0^\infty e^t d\mu(t) = \infty.
\end{aligned}
$$

In spite of this fact we obtain the following.

Theorem 2. 3.   *If $M \in BMO$, then $\mathcal{E}(M)$ is a uniformly integrable martingale.*

Proof.   As $\langle M \rangle_\infty < \infty$ a.s, it follows from (1.4) that $\mathcal{E}(M)_T > 0$ a.s for any stopping time T. Observe that

$$\mathcal{E}(M)_\infty / \mathcal{E}(M)_T = \exp\left\{(M_\infty - M_T) - (\langle M \rangle_\infty - \langle M \rangle_T)/2\right\}.$$

Then by the Jensen inequality we get

$$
\begin{aligned}
E[\mathcal{E}(M)_\infty | \mathcal{F}_T] &= E[\mathcal{E}(M)_\infty / \mathcal{E}(M)_T | \mathcal{F}_T] \mathcal{E}(M)_T \\
&\geq \exp\left(-\frac{1}{2} \| M \|_{BMO_2}^2\right) \mathcal{E}(M)_T,
\end{aligned}
$$

which implies that $\mathcal{E}(M) \in \mathcal{M}_u$. $\square$

In Chapter 3 we shall show that $M \mapsto \mathcal{E}(M) - 1$ is a continuous mapping of BMO into H_1.

Remark 2.3.   We proved in [35] that, if M is a right continuous BMO-martingale satisfying $\Delta M \geq -1 + \delta$ for some δ with $0 < \delta \leq 1$, then $\mathcal{E}(M)$ is a uniformly integrable martingale. However, the complete generalization to the right continuous BMO-martingales is impossible. In the following we state Sekiguchi's example [75]. Let $(\Omega, \mathcal{F})$ be the measureble space on $\mathbb{R}_+$ with Lebesgue measurable subsets, and let $P(d\omega) = e^{-\omega} d\omega, S(\omega) = \omega$ and $\mathcal{F}_t = \mathfrak{S}(S \wedge s; s \leq t)$ as in Example 2.1. Then each local martingale M starting at 0 on this probability system is represented by

$$M_t = f(s) \mathrm{I}_{\{s \leq t\}} - \int_0^{s \wedge t} f(s) ds$$

where f is a locally integrable function on $\mathbb{R}_+$. (see C.S. Chou and P.A. Meyer [4]). Furthermore, it is not difficult to verify that the exponential martingale $\mathcal{E}(M)$ is uniformly integrable if and only if

$$\lim_{t \to \infty} \int_0^t (1 + f(s)) ds = \infty.$$

Let now $f(s) = e^{-s} - 1$. Then $\int_0^t (1 + f(s)) ds = \int_0^t e^{-s} ds \leq 1$ clealy, so that $\mathcal{E}(M)$ is not uniformly integrable. On the other hand, since $\langle M \rangle_\infty = (e^{-s} - 1)^2 \leq 1$, if follows at once that $M \in BMO$.

Definition 2. 2.   Let $1 < p < \infty$. We say that $\mathcal{E}(M)$ satisfies (A_p) if

$$(2.10) \qquad \sup_T \left\| E[\{\mathcal{E}(M)_T / \mathcal{E}(M)_\infty\}^{1/(p-1)} | \mathcal{F}_T] \right\|_\infty < \infty,$$

where the supremum is taken over all stopping times T. Particularly, if $\sup_T \| \mathcal{E}(M)_T / \mathcal{E}(M)_\infty \|_\infty < \infty$, then we say that it satisfies (A_1).

From the Hölder inequality, it follows that if $1 \leq p < r$, then (A_p) implies (A_r).

The classical (A_p) condition has already appeared many times in the literature in connection with the weighted mean convergence of Fourier series. Let $0 \leq w \in L_1^{loc}(\mathbb{R}^n)$ and $1 < p < \infty$. As is well-known, B. Muckenhoupt proved in [62] that the inequality

$$\int_{R^n} f^*(x)^p w(x) dx \leq C_p \int_{R^n} |f(x)|^p w(x) dx,$$

where $f^*(x) = \sup_{x \in Q} \frac{1}{|Q|} \int_Q |f(y)| dy$, is valid for all $f \in L_p(w(x)dx)$ if and only if the weight function $w(x)$ satisfies the condition :

$$(2.11) \qquad \sup_Q \left(\frac{1}{|Q|} \int_Q w(x) dx \right) \left(\frac{1}{|Q|} \int_Q w(x)^{-1/(p-1)} dx \right)^{p-1} < \infty.$$

Here Q denotes a cube with sides parallel to the axes. Observe that the condition (2.10) is a probabilistic version of (2.11).

The next result is of fundamental importance in performing our investigation.

Theorem 2. 4. *The following conditions are equivalent.*

(a) $M \in BMO$.

(b) $\mathcal{E}(M)$ *satisfies* (A_p) *for some* $p \geq 1$.

(c) $\sup_T \left\| E \left[\log^+ \frac{\mathcal{E}(M)_T}{\mathcal{E}(M)_\infty} \Big| \mathcal{F}_T \right] \right\|_\infty < \infty$.

Proof. We first show that (a) implies (b). For that, take $p > 1$ such that $\| M \|_{BMO_2} < \sqrt{2}(\sqrt{p} - 1)$. Then, according to Theorem 2.1, we have

$$(2.12) \qquad E \left[\exp \left\{ \frac{1}{2(\sqrt{p}-1)^2} (\langle M \rangle_\infty - \langle M \rangle_T) \right\} \Big| \mathcal{F}_T \right] \leq C_p$$

for every stopping time T, with a constant $C_p > 0$. Let now $r = \sqrt{p} + 1$. Then the exponent conjugate to r is $s = (\sqrt{p} + 1)/\sqrt{p}$.
Since $\{s(\sqrt{p}-1)^2\}^{-1} - r(p-1)^{-2} = (p-1)^{-1}$, we have

$$\{\mathcal{E}(M)_T/\mathcal{E}(M)_\infty\}^{1/(p-1)}$$
$$= \exp \left\{ -\frac{1}{p-1}(M_\infty - M_T) - \frac{r}{2(p-1)^2}(\langle M \rangle_\infty - \langle M \rangle_T) \right\}$$
$$\cdot \exp \left\{ \frac{1}{2s(\sqrt{p}-1)^2}(\langle M \rangle_\infty - \langle M \rangle_T) \right\}.$$

We apply Hölder's inequality with the exponents r and s :

$$E[\{\mathcal{E}(M)_T/\mathcal{E}(M)_\infty\}^{1/(p-1)} | \mathcal{F}_T]$$
$$\leq E \left[\mathcal{E}\left(-\frac{r}{p-1} M \right)_\infty \Big/ \mathcal{E}\left(-\frac{r}{p-1} M \right)_T \Big| \mathcal{F}_T \right]^{1/r}$$
$$\cdot E \left[\exp \left\{ \frac{1}{2(\sqrt{p}-1)^2}(\langle M \rangle_\infty - \langle M \rangle_T) \right\} \Big| \mathcal{F}_t \right]^{1/s}.$$

By Theorem 2.3 the first conditional expectation on the right hand side is equal to 1. Furthermore, it follows from (2.9) that the second term is dominated by some constant C_p. This proves (b).

Next, we show that (b) implies (c). Assume that $\mathcal{E}(M)$ satisfies the (A_p) condition for $p > 1$. Then by the Jensen inequality we find

$$\exp \left\{ \frac{1}{p-1} E \left[\log^+ \frac{\mathcal{E}(M)_T}{\mathcal{E}(M)_\infty} \Big| \mathcal{F}_T \right] \right\} \leq E \left[\left\{ \frac{\mathcal{E}(M)_T}{\mathcal{E}(M)_\infty} \right\}^{1/(p-1)} \Big| \mathcal{F}_T \right] + 1 \leq C_p$$

for every stopping time T, with a constant $C_p > 1$ independent of T. Therefore, we have

$$E\left[\log^+ \left.\frac{\mathcal{E}(M)_T}{\mathcal{E}(M)_\infty}\right| \mathcal{F}_T\right] \leq (p-1)\log C_p.$$

Finally, we derive (a) from (c). The usual stopping argument enables us to assume that M is a uniformly integrable martingale. Then from (c) it follows that

$$
\begin{aligned}
E[\langle M\rangle_\infty - \langle M\rangle_T | \mathcal{F}_T] &= 2E[M_T - M_\infty + \frac{1}{2}(\langle M\rangle_\infty - \langle M\rangle_T)|\mathcal{F}_T] \\
&\leq 2E\left[\log^+ \left.\frac{\mathcal{E}(M)_T}{\mathcal{E}(M)_\infty}\right| \mathcal{F}_T\right] \leq C
\end{aligned}
$$

for every stopping time T, with a constant $C > 0$. Hence, M belongs to the class BMO, and the equivalence of (a), (b) and (c) is established. $\qquad\square$

Remark 2.3.   In the case where M is a right continuous local martingale satisfying $-1 < \triangle M \leq C$ for some constant $C > 0$, it is not difficult to prove that, if $\mathcal{E}(M)$ satisfies (A_p) for some $p > 1$, then M is a BMO-martingale. (see [43])

We end this section with a remark on a relation between the Muckenhoupt (A_p) condition and the probabilistic (A_p) condition.

Remark 2.4.   Let $D = \{z : |z| < 1\}$ be the unit disc in the complex plane and let $0 < w \in L_1(\partial D, dm)$ where $m(d\theta)$ denotes the normalized Lebesgue measure on ∂D. Then the Muckenhoupt (A_p) condition for the weight function w on ∂D is stated as follows :

$$\sup_I \left(\frac{1}{m(I)}\int_I w\,dm\right)\left(\frac{1}{m(I)}\int_I w^{-\frac{1}{p-1}}\,dm\right)^{p-1} < \infty$$

where the supremum is taken over all intervals $I \subset \partial D$. We may assume that $\int_{\partial D} w\,dm = 1$. Next, let $B = (B_t, \mathcal{F}_t)$ be the complex Brownian motion starting at 0 and let $\tau = \inf\{t : |B_t| = 1\}$. Then the positive martingale W given by $W_t = E[w(B_\tau)|\mathcal{F}_t]$ $(0 \leq t < \infty)$ satisfies $E[W_\infty] = 1$, and further from a celebrated theorem of Kakutani it follows that

$$W_t = \mathcal{P}w \cdot (B_t) \quad \text{on } \{t < \tau\}$$

where $\mathcal{P}w$ is the Poisson integral of w.

In this setting T. Sekiguchi and Y. Shiota ([76]) showed that W *satisfies the probabilistic* (A_p) *condition, then* w *satisfies the Muckenhoupt* (A_p) *condition.* We prove it below, barrowing their idea. Firstly, assume that W satisfies the probabilistic (A_p) condition. Then, from the strong Markov property and the Kakutani theorem it follows that

$$\sup_T \left\|\mathcal{P}w \cdot (B_T)\{\mathcal{P}[w^{-\frac{1}{p-1}}] \cdot (B_T)\}^{p-1}\right\|_\infty < \infty$$

where the supremum is taken over all stopping times T with $T < \tau$ *a.s.* This implies that

$$\sup_{z \in D} \mathcal{P}w \cdot (z) \left\{\mathcal{P}[w^{-\frac{1}{p-1}}] \cdot (z)\right\}^{p-1} < \infty.$$

Secondly, for each $z = re^{i\theta} \in D$ let I_z be an interval on ∂D such that z is the center of I_z and $m(I_z) = 1 - r$. Note that for any interval I on ∂D there is a point z in D satisfying $I = I_z$. From the definition of I_z it follows immediately that if $e^{i\eta} \in I_z$, then $|\eta - \theta| \le (1 - r)\pi$. Since $P(r, \theta - \eta) \ge K\frac{1+r}{1-r}$ for some constant $K > 0$, we get

$$
\begin{aligned}
\mathcal{P}(w) \cdot (z) &\ge \int_I w(\eta)P(r, \theta - \eta)m(d\eta) \\
&\ge K\frac{1+r}{1-r}\int_I w(\eta)m(d\eta)
\end{aligned}
$$

and in the same way

$$
\mathcal{P}[w^{-\frac{1}{p-1}}] \cdot (z) \ge K\frac{1+r}{1-r}\int_I w^{-\frac{1}{p-1}}(\eta)m(d\eta).
$$

Combining these two estimates shows that

$$
\begin{aligned}
&\left\{\frac{1}{m(I)}\int_I w(\eta)m(d\eta)\right\}\left\{\frac{1}{m(I)}\int_I w(\eta)^{-\frac{1}{p-1}}m(d\eta)\right\}^{p-1} \\
&\le \left\{\frac{1-r}{m(I)(1+r)K}\right\}^p \mathcal{P}w \cdot (z)\left\{\mathcal{P}\left[w^{-\frac{1}{p-1}}\right] \cdot (z)\right\}^{p-1} \\
&\le \left\{\frac{1}{K(1+r)}\right\}^p \mathcal{P}w \cdot (z)\left\{\mathcal{P}\left[w^{-\frac{1}{p-1}}\right] \cdot (z)\right\}^{p-1} \\
&\le K^{-p}\mathcal{P}w \cdot (z)\left\{\mathcal{P}\left[w^{-\frac{1}{p-1}}\right] \cdot (z)\right\}^{p-1}.
\end{aligned}
$$

The last expression is bounded by some constant C_p depending only on p, because W satisfies (A_p). Thus w satisfies the Muckenhoupt (A_p) condition. However, the converse is not true. Precisely speaking, there is a weight function w on ∂D such that, although it satisfies the classicl (A_p) condition for some $p > 1$, the weight martingale $W = w(B^\tau)$ corresponding to w satisfies no (A_p). For example, let $w(t) = |t|^\lambda$ where $1 < \lambda < \infty$. Then w satisfies (A_p) for $p > 1 + \lambda$ (see [77] and [62]). On the other hand, let $p > 1 + \lambda$. Observe that $P(r, \eta) \ge \frac{1-r}{1+r}$. If $z = re^{i\theta} \in D$, then we have

$$
\begin{aligned}
&\mathcal{P}w \cdot (z)\{\mathcal{P}[w^{-\frac{1}{p-1}}] \cdot (z)\}^{p-1} \\
&= \int_0^{2\pi} \eta^\lambda P(r, \eta)m(d\eta)\left\{\int_0^{2\pi} \eta^{-\frac{\lambda}{p-1}}P(r, \eta)m(d\eta)\right\}^{p-1} \\
&\ge \frac{1-r}{1+r}\int_0^{2\pi} \eta^\lambda m(d\eta)\left\{\int_{0<\eta\le(1-r)\pi} \eta^{-\frac{\lambda}{p-1}}P(r, \eta)m(d\eta)\right\}^{p-1} \\
&\ge C_{p,\lambda}\frac{1-r}{1+r}\left\{\frac{1+r}{1-r}\int_{0<\eta\le(1-r)\pi} \eta^{-\frac{\lambda}{p-1}}d\eta\right\}^{p-1} \\
&\ge C_{p,\lambda}\left(\frac{1+r}{1-r}\right)^{p-2}\left\{\frac{2}{1-\frac{\lambda}{p-1}}(1-r)^{1-\frac{\lambda}{p-1}}\right\}^{p-1} \\
&\ge C_{p,\lambda}\frac{(1+r)^{p-2}}{(1-r)^{\lambda-1}} \longrightarrow \infty \quad \text{as } r \to 1 - 0,
\end{aligned}
$$

so W satisfies no (A_u). But the case $p = 2$ is an exception. Sekiguchi and Shiota showed that w satisfies (A_2) if and only if W satisfies (A_2). We shall sketch their proof.

In short, it suffices to verify that if w satisfies the Muckenhoupt (A_2) condition, then W satisfies the probabilistic (A_2) condition. To see this, let $f = \log w$, which is a BMO-function on ∂D. Then it is not difficult to show that

$$\sup_{z \in D} \exp(\mathcal{P}f \cdot (z))\mathcal{P}[w^{-1}] \cdot (z) < \infty.$$

Since w^{-1} also satisfies (A_2), we have

$$\sup_{z \in D} \exp(-\mathcal{P}f \cdot (z))\mathcal{P}w \cdot (z) < \infty.$$

Combining these estimates, we can obtain

$$\sup_{z \in D} \mathcal{P}w \cdot (z)\mathcal{P}[w^{-1}] \cdot (z) < \infty.$$

Thus W satisfies (A_2).

2.4   Fefferman's inequality

In this section we shall prove the well-known result that the space BMO can be identified with the dual space of II_1. The key to the proof is the next inequality which was originally obtained by C. Fefferman ([17]) in the classical analysis and in [21] by R. K. Getoor and M. J. Sharpe for continuous martingales. The generalization to right continuous martingales was established in [59] by P. A. Meyer. Fefferman's inequality plays an essential role in dealing with various questions about BMO-martingales.

Theorem 2. 5.   *If $X \in H_1$ and $Y \in BMO$, then*

$$(2.13) \qquad E\left[\int_0^\infty |d\langle X, Y\rangle_s|\right] \leq \sqrt{2}\| X \|_{H_1}\| Y \|_{BMO_2}.$$

Proof.   We recall briefly the proof due to Getoor and sharpe. Firstly, observe that the inquality

$$E\left[\int_0^\infty |d\langle X, Y\rangle_s|\right]^2 \leq E\left[\int_0^\infty \langle X\rangle_s^{-1/2} d\langle X\rangle_s\right] E\left[\int_0^\infty \langle X\rangle_s^{1/2} d\langle Y\rangle_s\right].$$

holds in general. The first expectation on the right hand side is smaller than $2\| X \|_{H_1}$. On the other hand, by the integration by parts, the second expectation is

$$\begin{aligned}
E\left[\langle X\rangle_\infty^{1/2}\langle Y\rangle_\infty - \int_0^\infty \langle Y\rangle_s d\langle X\rangle_s^{1/2}\right] &= E\left[\int_0^\infty (\langle Y\rangle_\infty - \langle Y\rangle_s)d\langle X\rangle_s^{1/2}\right] \\
&= E\left[\int_0^\infty E[\langle Y\rangle_\infty - \langle Y\rangle_s|\mathcal{F}_s]d\langle X\rangle_s^{1/2}\right] \\
&\leq \| Y \|_{BMO_2}^2 \| X \|_{H^1}.
\end{aligned}$$

Thus the theorem is proved.   □

Fefferman's inequality implies that $BMO \subset H_1^*$.

Theorem 2. 6.   *The dual of H_1 is BMO. Precisely speaking, if $f \in H_1^*$, then there exists a unique $M \in BMO$ such that $f(X) = E[\langle X, M\rangle_\infty]$ for every $X \in H_1$.*

Proof. We will adopt a very simple and direct proof due to Meyer ([60]) who established this duality theorem for right continuous martingales.

Let f be a continuous linear functional on H_1. Then

$$|f(X)| \leq c\| X \|_{H_1} \leq c\| X \|_{H_2}$$

by the Jensen inequality, and so f induces a continuous linear functional on H_2. From the duality theorem for H_2 it follows that there is an $M \in H^2$ such that $f(X) = E[X_\infty M_\infty]$ for all $X \in H_2$. Then $f(X) = E[\langle X, M \rangle_\infty]$, because $XM - \langle X, M \rangle$ is a uniformly integrable martingale. To show that $M \in BMO$, let T be a stopping time and let $U = I_A \cdot sgn(M_\infty - M_T)$ where $A \in \mathcal{F}_T$. Consider now the martingale N given by $N_t = E[U - E[U|\mathcal{F}_T]|\mathcal{F}_t]$ $(0 \leq t < \infty)$. It is clear that $|N_\infty| \leq 2I_A$, and so

(2.14) $$|f(N)| \leq c\| N \|_{H^2} \leq cP(A).$$

On the other hand, we have

$$
\begin{aligned}
f(N) &= E[N_\infty M_\infty] \\
&= E[(U - E[U|\mathcal{F}_T])M_\infty] \\
&= E[U(M_\infty - M_T)] \\
&= E[|M_\infty - M_T| : A].
\end{aligned}
$$

Combining this with (2.14) gives

$$E[|M_\infty - M_T| : A] \leq cP(A) \quad (A \in \mathcal{F}_T),$$

which implies that $\|M\|_{BMO_1} \leq c$.

Let now $X \in H_1$. Since H_2 is dense in H_1, there exists a sequence $\{X^{(n)}\}_{n=1,2,\cdots}$ where $X^{(n)} \in H_2$ for every n such that

$$X^{(n)} \to X \quad \text{in } H^1 \quad (n \to \infty)$$

Then, as $n \to \infty$

$$|E[\langle X^{(n)}, M \rangle_\infty] - E[\langle X, M \rangle_\infty]| \leq \sqrt{2}\| X^{(n)} - X \|_{H_1}\| M \|_{BMO_2} \to 0,$$

so that we have

$$f(X) = \lim_{n \to \infty} f(X^{(n)}) = \lim_{n \to \infty} E[\langle X^{(n)}, M \rangle_\infty] = E[\langle X, M \rangle_\infty].$$

Thus the proof is complete. $\qquad\qquad\square$

Remark 2.5. It should be noted that $X_\infty M_\infty$ is not always integrable for $X \in H_1$ and $M \in BMO$. We give below an example. Let $D = \{z : |z| < 1\}$ be the unit disc in the complex plane, and let

$$f(e^{it}) = \begin{cases} (t\log^2 |t|)^{-1} & \text{if } |t| \leq \frac{1}{2} \\ 0 & \text{if } \frac{1}{2} < |t| \leq \pi \end{cases}$$

$$g(e^{it}) = \begin{cases} \log |t| & \text{if } |t| \leq 1 \\ 0 & \text{if } 1 < |t| \leq \pi \end{cases}$$

Note that $fg \notin L_1(dm)$ where $m(dt)$ is the normalized Lebesgue measure on ∂D. Now, let $B = (B_t, \mathcal{F}_t)$ be the complex Brownian motion starting at 0, and let $\tau = \inf\{t : |B_t| = 1\}$. Then the martingale $X = \mathcal{P}f \cdot (B^\tau)$ belongs to H_1, where $\mathcal{P}f$ is the Poisson integral of f. On the other hand, $M = \mathcal{P}g \cdot (B^\tau)$ is a BMO-martingale, because g is a BMO-function. However $X_\infty M_\infty = f(B^\tau)g(B^\tau)$, which is not integrable.

Refer to [30] or [12] for further discussion about the expression of the duality between H_1 and BMO.

Theorem 2. 7.   *Let X be a local martingale. Then we have*

(2.15) $\| X \|_{H_1} \leq \sup\{E[\langle X, Y\rangle_\infty] : \| Y \|_{BMO_2} \leq 1\}$

(2.16) $\| X \|_{BMO_2} \leq \sup\{E[\langle X, Y\rangle_\infty] : \| Y \|_{H_1} \leq 1\}$

Proof.   We begin with the proof of (2.15). Let (T_n) be a non-decreasing sequence of stopping times with $\lim_{n\to\infty} T_n = \infty$ a.s such that $X^{T_n} \in H_1$ for every n. It is clear that $\langle X^{T_n}, Y\rangle = \langle X, Y^{T_n}\rangle$, $\| Y^{T_n} \|_{BMO_2} \leq \| Y \|_{BMO_2}$ and $\lim_{n\to\infty} \| X^{T_n} \|_{H_1} = \| X \|_{H_1}$. Therefore, we may assume that $X \in H_1$. Let now ε be an arbitrary positive real number, and define $Y_t = \int_0^t C_s dX_s$, where $C_t = E[(\varepsilon + \langle X\rangle_\infty)^{-1/2}|\mathcal{F}_t]$. Then we can easily verify that $\| Y \|_{BMO_2} \leq 1$. Furthermore, $\langle X\rangle$ being continuous, we have

$$E[\langle X, Y\rangle_\infty] = E\left[\int_0^\infty C_s d\langle X\rangle_s\right]$$
$$= E[(\varepsilon + \langle X\rangle_\infty)^{-1/2}\langle X\rangle_\infty],$$

which increases to $\| X \|_{H_1}$ as $\varepsilon \to 0$.

Next, we shall give the proof of (2.16), following the idea of Meyer. Let us denote by d its right hand side, and let T be any stopping time. It is sufficient to show that the inequality

$$E[\langle X\rangle_\infty - \langle X\rangle_T : A] \leq d^2 P(A)$$

hold for every $A \in \mathcal{F}_T$. For simplicity, set $U = \langle X\rangle_\infty - \langle X\rangle_T$. The stopping time argument enables us to assume that $X \in BMO$, and so $E[UI_A] < \infty$. The process D given by $D_t = I_{A\cap\{T<t\}}$ is predictable such that $D^2 = D$. Then we have

$$\langle D \circ X, X\rangle_\infty = \langle D \circ X\rangle_\infty = UI_A,$$

form which it follows that

$$E[UI_A] \leq d\| D \circ X \|_{H^1} = dE\left[I_A\sqrt{UI_A}\right].$$

The right hand side is smaller than $dP(A)^{1/2}E[UI_A]^{1/2}$ by the Schwarz inequality. Consequently we get $E[UI_A] \leq d^2 P(A)$, which completes the proof.   □

2.5   The Garnett-Jones theorem

The aim of this section is to give comparable upper and lower bounds for the distance in BMO to L_∞. For each $M \in BMO$, let $a(M)$ be the supremum of the set of a for which

$$\sup_{T} \| E[\exp(a|M_\infty - M_T|)|\mathcal{F}_T] \|_\infty < \infty,$$

and let $d_p(\ ,\ )$ be the distance on the space BMO deduced from the norm $\| \cdot \|_{BMO_p}$ by the usual procedure. Then there is a very beautiful relation between $a(M)$ and $d_1(M, L_\infty)$ as follows :

Theorem 2. 8 (N. Th. Varopoulos [83], M. Emery [14]). *For $M \in BMO$ we have*

(2.17)
$$\frac{1}{4d_1(M, L_\infty)} \leq a(M) \leq \frac{4}{d_1(M, L_\infty)}$$

This result was originally obtained in 1978 by J. Garnett and P. Jones ([18]) in classical analysis. Its probabilistic version was first established in 1980 by N. Th. Varopoulos for Brownian martingales and in 1985 by M. Emery for continuous martingales. The assumption on the sample continuity of all martingale is essential for the validity of the right-hand side inequaliaty, but the left-hand side holds without it. The reader is referred to Emery([14]) for the details.

To prove the right-hand side, we need the following.

Lemma 2. 2 (N. Th. Varopoulos [83]). *Let S and T be stopping times such that $S \leq T$ a.s., and assume that*

(2.18)
$$P(T < \infty|\mathcal{F}_S) \leq c^m$$

where $0 < c < 1$ and $m \in \mathbb{N}$.
Then there exists a sequence $\{R_i\}_{i=0,1,\cdots,m}$ of stopping times such that

(2.19)
$$S = R_0 \leq R_1 \leq \cdots \leq R_m = T$$

and for each $j = 0, 1, \cdots, m-1$

(2.20)
$$P(R_{j+1} < \infty|\mathcal{F}_{R_j}) \leq c.$$

Proof. Let $X_t = P(T < \infty|\mathcal{F}_t)$ $(0 \leq t < \infty)$ and let us define the stopping times

$$S_j = \inf\{t \geq S : X_t \geq c^{m-j}\} \quad (j = 1, 2, \cdots, m).$$

It is easy to see that $S \leq S_0 \leq S_1 \leq \cdots \leq S_m \leq T$. Let now $R_0 = S, R_j = S_j$ $(1 \leq j \leq m-1)$ and $R_m = T$. The claim is that the stopping times $R_0, R_1, \cdots, R_m$ satisfy the conditions of the lemma. Note that $X_S \leq c^m$ by the assumption and $X_T = I_{\{T<\infty\}}$. On the other hand, from the defintion of the stopping times R_i it follows immediately that

$$c^{m-j-1}P(R_{j+1} < \infty|\mathcal{F}_{R_j}) \leq E[X_{R_{j+1}}|\mathcal{F}_{R_j}] = X_{R_j} \leq c^{m-j}.$$

Thus we obtain

$$P(R_{j+1} < \infty|\mathcal{F}_{R_j}) \leq c \quad (j = 1, 2, \cdots, m-1).$$

Further, observing that $\mathcal{F}_{R_0} \subset \mathcal{F}_{S_0}$ and $S_m \leq R_m$ it follows that

$$P(R_1 < \infty|\mathcal{F}_{R_0}) = E[P(S_1 < \infty|\mathcal{F}_{S_0})|\mathcal{F}_{R_0}] \leq c$$

and

$$P(R_m < \infty | \mathcal{F}_{R_{m-1}}) \le P(S_m < \infty | \mathcal{F}_{S_{m-1}}) \le c.$$

This completes the proof of Lemma.   □

We can now give the :

Proof of Theorem 2.8 :   The proof we will give below is due to M. Emery ([14]).
We begin with the proof of the inequality on the right. For that, let $M \in BMO$. It
suffices to verify that if $0 \le \alpha < a(M)$, then $\alpha \le 4/d_1(M, L_\infty)$. First of all let $k > 0$
be a constant such that the inequality

$$E[\exp(\alpha | M_\infty - M_T |) | \mathcal{F}_T] \le \exp(\alpha k)$$

is valid for every stopping tome T. Let $0 < \varepsilon < 1$, $m > \alpha k / \varepsilon^2$ and $\beta = m\varepsilon / \alpha$ where
m is an integer. Define now the stopping times T_n inductively by letting $T_0 = 0$ and

$$T_j = \inf\{t > T_{j-1} : |M_t - M_{T_{j-1}}| = \beta\}$$

for $j \ge 1$. From the definition of T_j it follows at once that

$$\{T_j < \infty\} \subset \left\{ \left| M_{T_j} - M_{T_{j-1}} \right| = \beta \right\},$$

so we have

$$
\begin{aligned}
e^{\alpha\beta} P(T_j < \infty | \mathcal{F}_{T_{j-1}}) &\le E[\exp(\alpha | M_{T_j} - M_{T_{j-1}} |) | \mathcal{F}_{T_{j-1}}] \\
&\le E[\exp(\alpha | M_\infty - M_{T_{j-1}} |) | \mathcal{F}_{T_{j-1}}] \\
&\le e^{\alpha k}
\end{aligned}
$$

If we set $c = e^{-\varepsilon(1-\varepsilon)}$, then the above computation shows that for every $j \ge 1$

$$P(T_j < \infty | \mathcal{F}_{T_{j-1}}) \le c^m.$$

Next, let us consider the increasing processes

$$A^+ = \sum_{j=1}^{\infty} I_{\{M_{T_j} - M_{T_{j-1}} = \beta\}} I_{[\![T_j, \infty[\![}$$

and

$$A^- = \sum_{j=1}^{\infty} I_{\{M_{T_j} - M_{T_{j-1}} = -\beta\}} I_{[\![T_j, \infty[\![},$$

where $[\![T_j, \infty[\![$ is the stochastic interval $\{(t, \omega) \in \mathbb{R}_+ \times \Omega : T_j(\omega) \le t < \infty\}$. Observing
that $\{M_{T_j} - M_{T_{j-1}} = \beta\} \in \mathcal{F}_{T_j}$, we can rewrite A^+ as follows :

$$A^+ = \sum_{j=1}^{\infty} I_{[\![S_j, \infty[\![}$$

where each S_j is a stopping time and $S_j \uparrow \infty$ a.s. Then, it is not difficult to see that

$$P(S_n < \infty | \mathcal{F}_{S_{n-1}}) \le c^m$$

for every $n \geq 1$. So, by Lemma 2.2 there exist stopping times R_k such that $R_0 = 0, R_n \uparrow \infty$ $a.s, R_{nm} = S_n$ and $P(R_j < \infty | \mathcal{F}_{R_{j-1}}) \leq c$ for every $j \geq 1$. Let now

$$B^+ = \frac{1}{m} \sum_{j=1}^{\infty} \mathrm{I}_{[R_j, \infty[},$$

which is clearly an increasing process, and let

$$N_t^+ = \beta E \left[B_\infty^+ | \mathcal{F}_t \right] \quad (0 \leq t < \infty).$$

From the definitions of A^+ and B^+ it follows that $A^+ \leq B^+ \leq A^+ + 1$. In the same way let us define the increasing process B^- and the martingale N^- corresponding to A^-, and we also have $A^- \leq B^- \leq A^- + 1$. Let $N = N^+ - N^-$. We are going to show that $L = M - N$ is a bounded martingale. Since $|A_\infty^\pm - B_\infty^\pm| \leq 1, N_\infty = \beta(B_\infty^+ - B_\infty^-)$, and $|M_t - M_{T_j}| \leq \beta$ for $T_j \leq t < T_{j+1}$. We have

$$
\begin{aligned}
L_\infty &= M_\infty - \beta(B_\infty^+ - B_\infty^-) \\
&= \{M_\infty - \beta(A_\infty^+ - A_\infty^-)\} + \beta(A_\infty^+ - B_\infty^+) - \beta(A_\infty^- - B_\infty^-)
\end{aligned}
$$

and

$$M_t - \beta(A_t^+ - A_t^-) = \sum_{j=1}^{\infty} (M_t - M_{T_j}) \mathrm{I}_{\{T_j \leq t < T_{j+1}\}} + M_t \mathrm{I}_{\{t < T_1\}},$$

from which it follows immediately that $|L| \leq 3\beta$. On the other hand, by the definition of B^+ we have

$$P \left(B_\infty^+ - B_{R_{j-1}}^+ > \frac{i}{m} \bigg| \mathcal{F}_{R_j} \right) = P(R_{j+i} < \infty | \mathcal{F}_{R_j}) \leq c^i.$$

This impies that

$$
\begin{aligned}
E[B_\infty^+ - B_{t-}^+ | \mathcal{F}_t] &= E[E[B_\infty^+ - B_{R_{j-1}}^+ | \mathcal{F}_{R_j}] | \mathcal{F}_t] \\
&\leq \frac{1}{m(1-c)}
\end{aligned}
$$

on the set $\{R_{j-1} < t \leq R_j\}$. So we get that $\| N^+ \|_{BMO_1} \leq \frac{2\beta}{m(1-c)}$. Combining this with the same estimation for N^- gives

$$\| N \|_{BMO_1} \leq \frac{4\beta}{m(1-c)}.$$

Observe that $N = M - L, L \in L_\infty, c = e^{-\varepsilon(1-\varepsilon)}$ and $\frac{\beta}{m} = \frac{\varepsilon}{\alpha}$. Then, we have

$$
\begin{aligned}
d_1(M, L_\infty) &\leq \| N \|_{BMO_1} \\
&\leq \frac{4\varepsilon}{\alpha(1 - e^{-\varepsilon(1-\varepsilon)})}.
\end{aligned}
$$

The last expression tends to $\frac{4}{\alpha}$ as $\varepsilon \to 0$, and so, letting $\alpha \uparrow a(M)$ gives that $d_1(M, L_\infty) \leq 4/a(M)$.

The proof of the inequality on the left is easy. Let $0 \le a < \dfrac{1}{4d_1(M, L_\infty)}$. Then $a \parallel M - N \parallel_{BMO_1} < \frac{1}{4}$ for some $N \in L_\infty$ and by the John-Nirenberg inequality (2.4)

$$E\left[\exp(a|M_{infty} - M_T|)|\mathcal{F}_T\right] \le e^{2a\|N\|_\infty} E\left[|\exp(a|(M - M)_\infty - (M - N)_T|)|\,\mathcal{F}_T\right]$$

$$\le \frac{e^{2a\|N\|_\infty}}{1 - 4a \parallel M - N \parallel_{BMO_1}}.$$

This implies that $a \le a(M)$. Thus the proof is complete. $\square$

Example 2.2.   Let B be a one dimensional Brownian motion starting at 0, and set $M_t = B_{t\wedge 1}$. Then $\langle M \rangle_\infty = 1$ clearly. Now we shall show that $M \in \bar{L}_\infty \setminus L_\infty$. It is easy to see that $M \notin L_\infty$, and further it follows from the Schwarz inequality that for every $\alpha > 0$

$$E[\exp\{\alpha(M_\infty - M_T)\}|\mathcal{F}_T]$$
$$= E[\exp\{\alpha(M_\infty - M_T) - \alpha^2(\langle M \rangle_\infty - \langle M \rangle_T)\}$$
$$\cdot \exp\{\alpha^2(\langle M \rangle_\infty - \langle M \rangle_T)\}|\mathcal{F}_T]$$
$$\le E[\mathcal{E}(2\alpha M)_\infty/\mathcal{E}(2\alpha M)_T|\mathcal{F}_T]^{\frac{1}{2}} E[\exp\{2\alpha^2(\langle M \rangle_\infty - \langle M \rangle_T)\}|\mathcal{F}_T]^{\frac{1}{2}}$$
$$\le \exp(\alpha^2),$$

where T is an arbitrary stopping time. The same argument works if M is replaced by $-M$, so that for every $\alpha > 0$

$$E[\exp\{\alpha|M_\infty - M_T|\}|\mathcal{F}_T] \le 2\exp(\alpha^2).$$

This implies that $a(M) = \infty$. Thus $M \in \bar{L}_\infty$ by Theorem 2.8.

Example 2.3   As is remarked in Section 2.1, there exists a BMO-martingale M such that $\exp(M_\infty)$ is not integrable. In such a case we have $a(M) \le 1$, and so using Theorem 2.8 shows that $d_1(M, L_\infty) \ge 1/4$. This is an instance where L_∞ is not dense in BMO.

C. Dellacherie, P. A. Meyer and M. Yor proved in [7] that L_∞ is neither closed nor dense in BMO whenever $BMO \ne L_\infty$. We shall explain it later. On the other hand, in the classical setting it was implicitly shown by J. Garnett and P. Jones ([18]) that a locally integrable function f on R^d belongs to the BMO-closure of L_∞ if and only if both e^f and e^{-f} satisfy the Muckenhoupt (A_p) condition for all $p > 1$. We first establish a probabilistic analogue of their result. For a uniformly integrable martingale M, let

$$p(M) = \inf\{p > 1 : E[\exp(M_\infty)|\mathcal{F}.], E[\exp(-M_\infty)|\mathcal{F}.] \quad \text{have } (A_p)\}.$$

From Hölder's inequality it follows that $E[\exp(M_\infty)|\mathcal{F}.]$ has (A_p) for every $p > p(M)$.

Lemma 2. 3 . *Let $1 < p < \infty$. Then the following conditions are equivalent.*

(α_p) $\displaystyle\sup_T \left\| E\left[\exp\left(\frac{1}{p-1}|M_\infty - M_T|\right)\Big|\mathcal{F}_T\right]\right\|_\infty < \infty.$

(β_p) $\displaystyle\sup_T \left\| E\left[\exp\left(\frac{1}{p-1}M_\infty\right)\Big|\mathcal{F}_T\right] E\left[\exp\left(-\frac{1}{p-1}M_\infty\right)\Big|\mathcal{F}_T\right]\right\|_\infty < \infty.$

Proof.   For any stopping time T we have

$$E\left[\exp\left(\frac{1}{p-1}M_\infty\right)\bigg|\mathcal{F}_T\right]E\left[\exp\left(-\frac{1}{p-1}M_\infty\right)\bigg|\mathcal{F}_T\right]$$

$$= E\left[\exp\left\{\frac{1}{p-1}(M_\infty - M_T)\right\}\bigg|\mathcal{F}_T\right]E\left[\exp\left\{-\frac{1}{p-1}(M_\infty - M_T)\right\}\bigg|\mathcal{F}_T\right]$$

$$\leq E\left[\exp\left(\frac{1}{p-1}|M_\infty - M_T|\right)\bigg|\mathcal{F}_T\right]^2.$$

Thus (α_p) implies (β_p).

On the other hand, since $\{\exp(-\frac{1}{p-1}M_t),\mathcal{F}_t\}$ is a submartingale, we have

$$E\left[\exp\left\{\frac{1}{p-1}(M_\infty - M_T)\right\}\bigg|\mathcal{F}_T\right]$$

$$= E\left[\exp\left(\frac{1}{p-1}M_\infty\right)\bigg|\mathcal{F}_T\right]\exp\left(-\frac{1}{p-1}M_T\right)$$

$$\leq E\left[\exp\left(\frac{1}{p-1}M_\infty\right)\bigg|\mathcal{F}_T\right]E\left[\exp\left(-\frac{1}{p-1}M_\infty\right)\bigg|\mathcal{F}_T\right].$$

The same argument works if M is replaced by $-M$. Then (α_p) follows at once from (β_p). □

Lemma 2. 4.   *If $p(M) < \infty$, then $p(M) \leq 2$, $M \in BMO$ and*

$$p(M) - 1 = \frac{1}{a(M)}.$$

Proof.   Firstly, let $p > p(M)$. From the definition of $p(M)$ it follows that both $E[\exp(M_\infty)|\mathcal{F}]$ and $E[\exp(-M_\infty)|\mathcal{F}]$ satisfy (A_p). On the other hand, by Hölder's inequality we have

$$1 \leq E[\exp(M_\infty)|\mathcal{F}_T]E[\exp(-M_\infty)|\mathcal{F}_T].$$

Therefore, the (A_p) conditions

$$E[\exp(M_\infty)|\mathcal{F}_T]E\left[\exp\left(-\frac{1}{p-1}M_\infty\right)\bigg|\mathcal{F}_T\right]^{p-1} \leq C_p$$

and

$$E[\exp(-M_\infty)|\mathcal{F}_T]E\left[\exp\left(\frac{1}{p-1}M_\infty\right)\bigg|\mathcal{F}_T\right]^{p-1} \leq C_p$$

yield (β_p). Then (α_p) holds by Lemma 2.3, from which it follows that $M \in BMO$ and $\frac{1}{p-1} \leq a(M)$. Next we show that if $p(M) < \infty$, then $p(M) \leq 2$.

By the definition of $p(M)$, both $E[\exp(M_\infty)|\mathcal{F}]$ and $E[\exp(-M_\infty)|\mathcal{F}]$ satisfy (A_p) for some $p > 1$, and so $E[\exp(\pm\frac{1}{p-1}M_\infty)|\mathcal{F}]$ has (A_q) where $\frac{1}{p} + \frac{1}{q} = 1$, because

$$E\left[\exp\left(\frac{1}{p-1}M_\infty\right)\bigg|\mathcal{F}_T\right]E\left[\exp\left\{-\frac{1}{q-1}\left(\frac{1}{p-1}M_\infty\right)\right\}\bigg|\mathcal{F}_T\right]^{q-1}$$

$$= E[\exp(-M_\infty)|\mathcal{F}_T]^{\frac{1}{p-1}}E\left[\exp(\frac{1}{p-1}M_\infty)\bigg|\mathcal{F}_T\right] \leq C_p,$$

as well as a similar result with M and $-M$ interchanged. In general, (A_q) implies $(A_{q-\varepsilon})$ for some $\varepsilon > 0$ with $0 < \varepsilon < q-1$, which will be proved in Section 2 of Chapter 3. Then we have $p(\frac{1}{p-1}M) < q$ and so

$$\frac{1}{a\left(\frac{1}{p-1}M\right)} \le p\left(\frac{1}{p-1}M\right) - 1 < q - 1.$$

As $a(\frac{1}{p-1}M) = (p-1)a(M)$, we find that $a(M) > 1$. This implies (α_2) and then we get (β_2) by Lemma 2.3, namely,

$$E[\exp(M_\infty)|\mathcal{F}_T]E[\exp(-M_\infty)|\mathcal{F}_T] \le C_2.$$

Thus the martingales $E[\exp(M_\infty)|\mathcal{F}.]$ and $E[\exp(-M_\infty)|\mathcal{F}.]$ have (A_2), so that $p(M) \le 2$.

Finally we shall show that $p(M) - 1 \le \frac{1}{a(M)}$ if $p(M) < \infty$. For that, it is enough to verify that if $p - 1 > \frac{1}{a(M)}$, then $p \ge p(M)$. If $p \ge 2$, it is trivial, because $p(M) \le 2$. So we let $1 < p < 2$ and $p - 1 > \frac{1}{a(M)}$. As $a(M) > \frac{1}{p-1}$, we get (α_p), that is, (β_p) is valid. Then, an application of the Jensen inequality yields

$$E[\exp(M_\infty)|\mathcal{F}_T]^{\frac{1}{p-1}} E\left[\exp\left(-\frac{1}{p-1}M_\infty\right)\Big|\mathcal{F}_T\right]$$
$$\le E\left[\exp\left(\frac{1}{p-1}M_\infty\right)\Big|\mathcal{F}_T\right] E\left[\exp\left(-\frac{1}{p-1}M_\infty\right)\Big|\mathcal{F}_T\right] \le C_p.$$

This means that $E[\exp(-M_\infty)|\mathcal{F}.]$ satisfies (A_p). In the same way we can verify that $E[\exp(M_\infty)|\mathcal{F}.]$ satisfies (A_p). Therefore, we find that $p(M) \le p$. $\square$

Theorem 2. 9.   *Let $M \in BMO$. Then M belongs to the BMO-closure of L_∞ if and only if both $E[\exp(M_\infty)|\mathcal{F}.]$ and $E[\exp(-M_\infty)|\mathcal{F}.]$ satisfy all (A_p).*

Proof.   Let $M \in BMO$. Then $p(M) < \infty$ clearly. Combining Theorem 2.8 with Lemma 2.4 shows that

$$\frac{1}{4}d_1(M, L_\infty) \le p(M) - 1 \le 4d_1(M, L_\infty),$$

from which it follows at once that $M \in \bar{L}_\infty$ if and only if $p(M) = 1$. This completes the proof. $\square$

Thus we have obtained a probabilistic version of the result given by Garnett and Jones. However, it seems to me that the above characterization of $\bar{L}_\infty$ is unsatisfactory. For it is an exponential martingale which plays an essential role in various questions concerning the absolute continuity of probability laws of stochastic processes, but neither $E[\exp(M_\infty)|\mathcal{F}.]$ nor $E[\exp(-M_\infty)|\mathcal{F}.]$ are exponential martingales. In Section 3.4 of Chapter 3 we shall give another characterization of $\bar{L}_\infty$ in the framework of exponential martingales.

As a corollary of Theorem 2.9, we can obtain the following.

Theorem 2. 10.   *In order that both $\mathcal{E}(M)$ and $\mathcal{E}(-M)$ satisfy all (A_p), a necessary and sufficient condition is that $E[\log \mathcal{E}(M)_\infty|\mathcal{F}.] \in \bar{L}_\infty$.*

Before proving the result given by Dellacherie, Meyer and Yor, we shall give two conditions equivalent to the assumption "$BMO \neq L_\infty$" under which they showed that L_∞ is neither closed nor dense in BMO.

Theorem 2. 11.   *The following conditions are equivalent.*

(a) $BMO = L_\infty$.
(b)   *Almost all sample functions of every martingale adapted to the underlying filtration $(\mathcal{F}_t)$ are constant.*
(c)   *The filtration $(\mathcal{F}_t)$ is constant, that is, $\mathcal{F}_t = \mathcal{F}_0$ for every $t \geq 0$*

Proof.   Firstly, we claim that (a) implies (b). Otherwise, there would be a continuous martingale M such that

$$P\{\omega \in \Omega : \text{the sample function } M_.(\omega) \text{ is not constant}\} > 0.$$

By using the usual stopping argument we may assume that $M \in BMO$. Suppose now $BMO = L_\infty$. Then the norms $\| X \|_{BMO_p}$ and $\| X \|_\infty$ on BMO are equivalent by the closed graph theorem. So there exists a constant $C > 0$, depending only on M, such that $\| K \circ M \|_\infty \leq C$ for any predictable process $K = (K_t, \mathcal{F}_t)$ with $|K| \leq 1$. On the other hand, there exist $t > 0$ and a partition $\Gamma : 0 = t_0 < t_1 < t_2 < \cdots < t_n = t$ of $[0, t]$ such that $P(A) > 0$ where $A = \{\sum_{i=1}^n |M_{t_i} - M_{t_{i-1}}| \geq 2C\}$. Let now $B_{i,1} = \{M_{t_i} - M_{t_{i-1}} \geq 0\}$ and $B_{i,-1} = \{M_{t_i} - M_{t_{i-1}} < 0\}$. Since $A = \bigcup_{\varepsilon_i = \pm 1, 1 \leq i \leq n} A \cap B_{1,\varepsilon_1} \cap \cdots \cap B_{n,\varepsilon_n}$, we have

$$P\left(A \cap B_{1,\varepsilon_1^*} \cap \cdots \cap B_{n,\varepsilon_n^*}\right) > 0,$$

for some $\varepsilon_i^*(i = 1, 2, \cdots, n)$. Let us consider the process K defined by

$$K_s = \sum_{i=1}^n \varepsilon_i^* I_{]t_{i-1}, t_i]}(s),$$

which is clearly a predictable process with $|K| \leq 1$. Thus $\|K \circ M\|_\infty \leq C$ must follow. On the contrary, we find

$$(K \circ M)_t = \sum_{i=1}^n |M_{t_i} - M_{t_{i-1}}| > 2C$$

on the set $A \cap B_{1,\varepsilon_1^*} \cap \cdots \cap B_{n,\varepsilon_n^*}$. Then, the negation of (b) causes a contradiction. Thus (a) implies (b).

Next, we claim that (b) implies (c). For any fixed $t > 0$, let $A \in \mathcal{F}_t$, and let us consider the martingale $M_s = P(A|\mathcal{F}_s) - P(A|\mathcal{F}_0)$ $(0 \leq s < \infty)$. Then $M_\infty = 0$ by (b) and so we have $I_A = P(A|\mathcal{F}_0)$. This means that $A \in \mathcal{F}_0$. Thus $\mathcal{F}_t = \mathcal{F}_0$.

Finally, in order to prove the implication (c)$\Longrightarrow$ (a), let $M \in BMO$. Then $E[|M_\infty| |\mathcal{F}_0] \in L_\infty$ by the definition of a BMO-martingale, and it follows immediately from (c) that $M_\infty \in L_\infty$. Thus we have $BMO = L_\infty$. This completes the proof.   □

Theorem 2. 12.   *If the filtration $(\mathcal{F}_t)$ is non-constant, then L_∞ is not closed in BMO.*

Proof. We shall prove that the contraposition is valid. If $H_1 = L_1$, then $BMO = L_\infty$ by the duality theorem, and so $(\mathcal{F}_t)$ is constant by Theorem 2.11. Thus it suffices to show that $H_1 = L_1$ whenever L_∞ is closed in BMO. If L_∞ is closed in BMO, then the norms $\|X\|_{BMO_p}$ and $\|X\|_\infty$ are equivalent on L_∞ by the closed graph theorem. On the other hand, from Theorem 2.7 it follows that

$$\|X\|_{H_1} \sim \sup\{E[X_\infty Y_\infty]; Y \in L_\infty, \|Y\|_{BMO_2} \le 1\}$$

and that

$$\|X\|_{H_1} \sim \sup\{E[X_\infty Y_\infty]; Y \in L_\infty, \|Y\|_\infty \le 1\} \sim \|X\|_1.$$

Consequently, we have $H_1 = L_1$.   □

Theorem 2. 13. *If $(\mathcal{F}_t)$ is non-constant, then L_∞ is not dense in BMO.*

In order to prove this, we need only the implication $(a) \Rightarrow (b)$ in the next lemma.

Lemma 2. 5. *Let $\mathcal{K} \subset H_1$. Then the following are equivalent.*
(a) *Every sequence in $\mathcal{K}$ has a subsequence which converges in the weak topology $\sigma(H_1, BMO)$.*
(b) *The family $\{X^*\}_{X \in \mathcal{K}}$ is uniformly integrable, where $X^* = \sup_{0 \le t < \infty} |X_t|$.*

Proof. We argue by contradiction that (a) implies (b). Let (a) be valid. Then $\sup_{X \in \mathcal{K}} E[X^*] < \infty$ by the Banach-Steinhaus theorem. Suppose first that the family $\{X^*\}_{X \in \mathcal{K}}$ is not uniformly integrable. Then it is not uniformly absolutely continuous, that is, there exist some $\varepsilon_0 > 0$ and $A_n \in \mathcal{F}, n = 1, 2, \cdots$, such that $P(A_n) \le 2^{-n}$ and

$$\sup_{X \in \mathcal{K}} \int_{A_n} X^* dP \ge \varepsilon_0.$$

Let now $B_n = \bigcup_{k=n}^\infty A_k$. It is easy to see that $B_1 \supset B_2 \supset \cdots, P(B_n) \le 2^{-n+1}$ and

$$\sup_{X \in \mathcal{K}} \int_{B_n} X^* dP \ge \varepsilon_0 \quad (n = 1, 2, \cdots).$$

Then

$$\int_{B_1} X^{(1)*} dP > \frac{1}{2}\varepsilon_0$$

for some $X^{(1)} \in \mathcal{K}$, and further, as $P(B_n) \to 0$, we have

$$\int_{B_1 \setminus B_{i_1}} X^{(1)*} dP > \frac{1}{2}\varepsilon_0$$

for some $i_1 \in \mathbb{N}$. By the same reason there exist $X^{(2)} \in \mathcal{K}$ and $i_2 > i_1$ such that

$$\int_{B_{i_1+1}} X^{(2)*} dP > \frac{1}{2}\varepsilon_0, \quad \int_{B_{i_1+1} \setminus B_{i_2}} X^{(2)*} dP > \frac{1}{2}\varepsilon_0.$$

In this way we can choose $X^{(1)}, X^{(2)}, \cdots$ in $\mathcal{K}$ and $i_0 = 1 < i_1 < i_2 < \cdots$ in $\mathbb{N}$ such that

$$\int_{B_{i_{n-1}+1}} X^{(n)*} dP > \frac{1}{2}\varepsilon_0, \quad \int_{B_{i_{n-1}+1} \setminus B_{i_n}} X^{(n)*} dP > \frac{1}{2}\varepsilon_0.$$

Let $C_n = B_{i_{n-1}+1} \setminus B_{i_n}$. It is clear that $C_n \cap C_m = \emptyset$ $(m \neq n)$ and for every $n \geq 1$

$$\int_{C_n} X^{(n)^*} dP > \frac{1}{2}\varepsilon_0.$$

By the Section Theorem there exists a positive random variable S_n defined on C_n such that

$$\int_{C_n} |X_{S_n}^{(n)}| dP > \frac{1}{3}\varepsilon_0.$$

If we let $S = \sum_{n=1}^\infty S_n I_{C_n}$, then

$$\int_{C_n} |X_S^{(n)}| dP > \frac{1}{3}\varepsilon_0 \quad (n = 1, 2, \cdots).$$

Thus the family $\{X_S^{(n)}\}_{n=1,2,\cdots}$ is not uniformly integrable.

However, by (a) there exists a subsequence $\{X^{(n_k)}\}_{k=1,2,\cdots}$ which converges in the weak topology $\sigma(H_1, BMO)$. Then it is not difficult to see that the sequence $\{X_S^{(n_k)}\}$ converges in the weak topology $\sigma(L_1, L_\infty)$.

Therefore, $\{X_S^{(n_k)}\}_{k=1,2,\cdots}$ is uniformly integrable by Dunford-Pettis Weak Compactness Criterion and our initial assumption has led to a contradiction.

For the proof of the implication (b)$\Longrightarrow$(a), see [7]. $\square$

Proof of Theorem 2.13 :   To prove that the contraposition is valid, suppose that L_∞ is dense in BMO. For the same reason as in the proof of Theorem 2.12 it suffices to show that $H_1 = L_1$, that is, $\|X\|_{H_1} \leq C\|X\|_1$ for $X \in H_1$. This means that if $X^{(n)} \in H_1$ and $X^{(n)}$ converges in L_1 to 0, then $X^{(n)}$ converges in H_1 to 0. To see this, we may assume that $\|X^{(n)}\|_{H_1} \leq 1$ for every $n \geq 1$. Let now $Y \in BMO$. Since $\bar{L}_\infty = BMO$ by the assumption, there exists a sequence $\{Y^{(n)}\}$ in L_∞ which converges in the space BMO to Y. From Fefferman's inequality it follows that

$$
\begin{aligned}
|E[\langle X^{(n)}, Y \rangle_\infty]| &\leq |E[\langle X^{(n)}, Y^{(k)} \rangle_\infty]| + |E[\langle X^{(n)}, Y - Y^{(k)} \rangle_\infty]| \\
&\leq |E[X_\infty^{(n)} Y_\infty^{(k)}]| + \sqrt{2}\|X^{(n)}\|_{H_1} \|Y^{(k)} - Y\|_{BMO_2} \\
&\leq |E[X_\infty^{(n)} Y_\infty^{(k)}]| + \sqrt{2}\|Y^{(k)} - Y\|_{BMO_2}.
\end{aligned}
$$

Then we have

$$\limsup_{n \to \infty} |E[\langle X^{(n)}, Y \rangle_\infty]| \leq \sqrt{2}\|Y^{(k)} - Y\|_{BMO_2} \longrightarrow 0 \quad (k \to \infty),$$

which implies that $X^{(n)} \longrightarrow 0$ in the weak topology $\sigma(H_1, BMO)$. Thus $\{X^{(n)^*}\}_{n=1,2,\cdots}$ is uniformly integrable by Lemma 2.5. Since $X_\infty^{(n)} \longrightarrow 0$ in L_1 $(n \to \infty)$, $X^{(n)^*}$ converges in probability to 0 by the Doob inequality. Consequently, $X^{(n)^*} \longrightarrow 0$ in L_1, namely, $X^{(n)} \longrightarrow 0$ in H_1 as $n \to \infty$. $\square$

2.6   The class H_∞

Recall that H_∞ denotes the class of all martingales with bounded quadratic variation. Obviously, H_∞ is contained in BMO and, as is remarked before, there is not always an inclusion relation between H_∞ and L_∞. However, we have the following.

Theorem 2. 14. $H_\infty \subset \bar{L}_\infty$

Proof.   If $M \in H_\infty$, then for every $\lambda > 0$

$$E[\exp\{\lambda(M_\infty - M_T)\}|\mathcal{F}_T]$$

$$= E\left[\left\{\frac{\mathcal{E}(2\lambda M)_\infty}{\mathcal{E}(2\lambda M)_T}\right\}^{\frac{1}{2}} \exp\{\lambda^2(\langle M\rangle_\infty - \langle M\rangle_T)\}\bigg|\mathcal{F}_T\right]$$

$$\leq E\left[\frac{\mathcal{E}(2\lambda M)_\infty}{\mathcal{E}(2\lambda M)_T}\bigg|\mathcal{F}_T\right]^{\frac{1}{2}} E[\exp\{2\lambda^2(\langle M\rangle_\infty - \langle M\rangle_T)\}|\mathcal{F}_T]^{\frac{1}{2}}$$

$$\leq E[\exp\{2\lambda^2(\langle M\rangle_\infty - \langle M\rangle_T)\}|\mathcal{F}_T]^{\frac{1}{2}} \leq C_\lambda.$$

The same argument works if M is replaced by $-M$ and we get

$$E[\exp\{\lambda(M_\infty - M_T)\}|\mathcal{F}_T] \leq E[\exp\{2\lambda^2(\langle M\rangle_\infty - \langle M\rangle_T)\}|\mathcal{F}_T]^{\frac{1}{2}}.$$

Therefore, from Theorem 2.8 it follows immediately that $M \in \bar{L}_\infty$.   □

Let now $b(M)$ denote the supremum of the set of b for which

$$\sup_T \left\|E[\exp\{b^2(\langle M\rangle_\infty - \langle M\rangle_T)\}|\mathcal{F}_T]\right\|_\infty < \infty.$$

Note that $BMO = \{M : b(M) > 0\}$ by Theorem 2.2.

Lemma 2. 6.   *Let* $M \in BMO$. *Then*

$$\frac{1}{\sqrt{2}d_2(M, H_\infty)} \leq b(M).$$

Proof.   Let $M \in BMO$ and let $0 < b < \dfrac{1}{\sqrt{2}d_2(M, H_\infty)}$. Then for some $N \in H_\infty$

$$b < \frac{1}{\sqrt{2}\|M - N\|_{BMO_2}}.$$

Since $\langle N\rangle_\infty \leq C$ for some constant $C \geq 0$, we find for $s < t$

$$\langle M\rangle_t - \langle M\rangle_s \leq 2(\langle M - N\rangle_t - \langle M - N\rangle_s) + 2C.$$

Then an application of Theorem 2.2 gives

$$E[\exp\{b^2(\langle M\rangle_\infty - \langle M\rangle_T)\}|\mathcal{F}_T]$$

$$\leq \exp(2b^2 C)E[\exp\{2b^2(\langle M - N\rangle_\infty - \langle M - N\rangle_T)\}|\mathcal{F}_T]$$

$$\leq \frac{\exp(2b^2 C)}{1 - 2b^2\|M - N\|_{BMO_2}^2}.$$

This implies that $b \leq b(M)$. Thus the lemma is proved.   □

I could not settle the question whether or not an inequality in the other direction

$$b(M) \leq \frac{C}{d_2(M, H_\infty)}$$

is valid. If this is true, then we can obtain a remarkable characterization of $\bar{H}_\infty$, that is, $\bar{H}_\infty = \{M : b(M) = \infty\}$. In this connection, it is significant to investigate whether or not $\bar{H}_\infty = \bar{L}_\infty$. In 1978 Dellacherie, Meyer and Yor ([7]) conjectured that H_∞ must be dense in BMO, but after two years I. V. Pavlov gave in [67] a counterexample. However, combining Theorems 2.13 and 2.14 shows that H_∞ is not dense in BMO except a trivial case. In parentheses, let $B = (B_t)$ be a one dimensional Brownian motion with $B_0 = 0$, and let $\tau = \inf\{t : |B_t| = 1\}$. Recall that $\exp(\frac{\pi^2}{8}\tau) \notin L_1$ by Lemma 1.3 in Chapter 1. Then we have $d_2(B^\tau, H_\infty) \geq \frac{2}{\pi}$ by Lemma 2.6. We are now going to show that $L_\infty \setminus \bar{H}_\infty \neq \emptyset$ under a very weak assumption.

Definition 2. 3.   *A stopping time T is said to be an innovation time if there exists a continuous local martingale X such that $\langle X \rangle_t < \langle X \rangle_T$ on $\{t < T\}$.*

The definition of an innovation time is introduced in [16] by M. Emery, C. Stricker and J. A. Yan. It is easy to see that if $(\mathcal{F}_t)$ is the Brownian filtration, then there exists an innovation time. The aim of this section is to prove the following.

Theorem 2. 15.   *Suppose the existence of a predictable innovation time T such that $P(T > 0) > 0$. Then there exists a bounded martingale which does not belong to the closure $\bar{H}_\infty$ in BMO.*

Throughtout this survey, we suppose that any local martingale adapted to the filtration $(\mathcal{F}_t)$ is continuous. Thus, any stopping time is predictable. In order to prove this result, we need the next three lemmas.

Lemma 2. 7.   *If there exists a continuous local martingale X such that $\langle X \rangle_\infty = \infty$ a.s., then there exists a bounded martingale which does not belong to $\bar{H}_\infty$.*

Proof.   Let $\theta_t = \inf\{s : \langle X \rangle_s > t\}$ and let $W_t = X_{\theta_t}$. Since $\langle X \rangle_\infty = \infty$ by the assumption, the process $W = (W_t, \mathcal{F}_{\theta_t})$ is a one dimensional Brownian motion as is well known. Next, let $\sigma = \inf\{t : |W_t| = 1\}$, which is clearly a stopping time with respect to the new filtration $(\mathcal{F}_{\theta_t})$. Note that $\exp(\frac{\pi^2}{8}\sigma)$ is not integrable. It is not difficult to verify that θ_σ is a stopping time with respect to $(\mathcal{F}_t)$. Consider now the process M given by $M_t = X_{t \wedge \theta_\sigma}$, which is a continuous local martingale over $(\mathcal{F}_t)$. Since X is constant on the stochastic interval $[\![t, \theta_{\langle X \rangle_t}]\!]$, we find

$$M_t = X_{\theta_{\langle X \rangle_t} \wedge \theta_\sigma} = X_{\theta_{\langle X \rangle_t \wedge \sigma}} = W_{\langle X \rangle_t \wedge \sigma},$$

from which it follows that $|M| \leq 1$. Furthermore, since $\langle M \rangle_\infty = \langle W \rangle_\sigma = \sigma$, $\exp(\frac{\pi^2}{8}\langle M \rangle_\infty)$ is not integrable. This implies that $d_2(M, H_\infty) \geq \frac{2}{\pi}$. Thus the lemma is poved. □

Lemma 2. 8.   *Let $A = (A_t, \mathcal{F}_t)$ be a right continuous increasing process such that $A_t < A_\infty$ for every finite t. Then there exists a positive continuous increasing process $C = (C_t, \mathcal{F}_t)$ such that $\int_0^\infty C_s dA_s = \infty$.*

Proof.   We shall prove it, following the idea of W.A.Zheng. Let $T_0 = 0$ and for $n = 0, 1, 2, \cdots$ let $T'_n = T_n + 1, T''_n = T_n + 2$ and

$$T_{n+1} = \inf\{t \geq T''_n : A_t > A_{T''_n}\}.$$

It is clear that $T_{n+1} \geq T''_n, T_{n+1} - T_n \geq 2$ and so $T_n \uparrow \infty$ a.s. Furthermore, we have $T_n < \infty$ a.s. for every n. In fact, if $T_k(\omega) < T_{k+1}(\omega) = \infty$ for some k, then $T''_k(\omega) < \infty$ and $A_{T''_k}(\omega) = A_\infty(\omega)$ by the definition of T_{k+1}. This is a contradiction. From the definition of T_{n+1} it follows that $A_t > A_{T''_n}$ for some t with $T''_n < t < T'_{n+1}$, so $A_{T''_n} < A_{T'_{n+1}-}$. Let us choose $\delta_n > 0$ so that $P(A_{T'_{n+1}-} - A_{T''_n} < \delta_n) < 2^{-n}$. we may assume that $\delta_n \downarrow 0$. Let now $C = (C_t, \mathcal{F}_t)$ be a continuous increasing process such that

$$
C = \begin{cases} 0 & \text{on } [0,1] \\ \text{linear} & \text{on } [T_n, T''_n] \\ \dfrac{1}{\delta_n} & \text{on } [T''_n, T_{n+1}] \end{cases}
$$

An application of the first Borel-Cantelli lemma gives

$$
P(\limsup\{A_{T'_{n+1}-} - A_{T''_n} < \delta_n\}) = 0,
$$

namely, for almost all ω, there exists an integer n_0 such that for $n \geq n_0$

$$
A_{T'_{n+1}-}(\omega) - A_{T''_n}(\omega) \geq \delta_n.
$$

Then we have

$$
\int_0^\infty C_t dA_t \geq \sum_{n=n_0}^\infty \int_{T''_n}^{T'_{n+1}} C_t dA_t \geq \sum_{n=n_0}^\infty \frac{1}{\delta_n}(A_{T'_{n+1}-} - A_{T''_n}) = \infty.
$$

Thus the lemma is proved.   □

Lemma 2. 9.   *If there exists an innovation time $T > 0$ a.s, then there exists a local martingale M satisfying $\langle M \rangle_\infty = \infty$ a.s.*

Proof.   Since we suppose that any martingale adapted to the filtration $(\mathcal{F}_t)$ is continuous, any stopping time is predictable. By the definition of an innovation time T, for some continuous local martingale X we have $\langle X \rangle_t < \langle X \rangle_T$ on $\{t < T\}$ and, T being predictable, there is a sequence $\{T_n\}$ of stopping times such that $T_0 = 0, T_n \uparrow T$ a.s and $T_n < T$ for every n. Let now $g_n : [n-1, n[\longrightarrow [0, \infty[$ be an increasing homeomorphic function, and for each n we set

$$
\tau_t = \max[T_{n-1}, \min\{T_n, g_n(t)\}] \quad (n-1 \leq t < n).
$$

Then $(\tau_t)_{0 \leq t < \infty}$ is a continuous change of time such that $\tau_0 = 0, \tau_n = T_n$ $(n = 1, 2, \cdots)$ and further $\tau_t < T$ for every finite t. So, the process Y defined by $Y_t = X_{\tau_t}$ $(0 \leq t < \infty)$ is continuous local martingale over the new filtration $(\mathcal{F}_{\tau_t})$, and for every finite t we have

$$
\langle Y \rangle_t = \langle X \rangle_{\tau_t} < \langle X \rangle_T = \langle Y \rangle_\infty.
$$

Thus from Lemma 2.8 it follows that $\int_0^\infty C_t d\langle Y \rangle_t = \infty$ a.s for some positive continuous increasing process $C = (C_t, \mathcal{F}_{\tau_t})$. Next, let $\sigma_t = \inf\{s : \tau_s > t\}$ and let $D_t = C_{\sigma_t}$. As is easily verified, each σ_t is a stopping time with respect to the filtration $(\mathcal{F}_{\tau_t})$, so that D_t is $\mathcal{F}_{\tau_{\sigma_t}}$-measurable. However, D_t is in fact $\mathcal{F}_t$-measurable, because $\tau_{\sigma_t} \leq t$ by the definition of σ_t. This guarantees that the stochastic integral $L_t = \int_0^t \sqrt{D_s} dX_s$ is well-defined. Since $L_{\tau_t} = \int_0^t \sqrt{D_{\tau_s}} dY_s$, we find

$$\langle L \rangle_\infty \geq \langle L \rangle_{\tau_\infty} = \int_0^\infty D_{\tau_s} d\langle Y \rangle_s = \int_0^\infty C_{\sigma_{\tau_s}} d\langle Y \rangle_s.$$

Noticing that C is increasing and $\sigma_{\tau_s} \geq s$, we have in conclusion

$$\langle L \rangle_\infty \geq \int_0^\infty C_s d\langle Y \rangle_s = \infty \ a.s.,$$

which completes the proof. $\qquad\qquad\square$

Proof of Theorem 2.15 :   Let T be an innovation time such that $P(T > 0) > 0$, and we set

$$\Omega' = \{T > 0\}, \quad \mathcal{F}_t' = \mathcal{F}_t|_{\Omega'}, \quad dP' = \frac{1}{P(\Omega')} I_{\Omega'} dP.$$

It is obvious that $\Omega' \in \mathcal{F}_0$, so that if X is a martingale, then $X I_{\Omega'}$ is a martingale under P' and in addition we have $\langle X I_{\Omega'} \rangle = \langle X \rangle I_{\Omega'}$. Conversely, if X' is a martingale over $(\mathcal{F}_t')$ under P', then there exists a martingale $X = (X_t, \mathcal{F}_t)$ under P such that $X' = X I_{\Omega'}$. So, X' is continuous and T is also an innovation time over $(\mathcal{F}_t')$. As $P'(T > 0) = 1$, from Lemmas 2.7 and 2.9 it follows that on the probability system $(\Omega', \mathcal{F}', P' : (\mathcal{F}_t'))$ there exists a bounded martingale M' for which $\exp(\frac{\pi^2}{8} \langle M' \rangle_\infty) \notin L_1(P')$. Let now $M = M' I_{\Omega'}$. Then it is a bounded martingale over $(\mathcal{F}_t)$ such that $\exp(\frac{\pi^2}{8} \langle M \rangle_\infty) \notin L_1(P)$. Thus $M \in L_\infty \setminus \bar{H}_\infty$. $\qquad\square$

As a corollary to Theorem 2.15 we shall show that a change of law gives sometimes rise to a morbid phenomenon. Let M be a martingale, and assume that the associated exponential process $\mathcal{E}(M)$ is a uniformly integrable martingale. This means that $d\hat{P} = \mathcal{E}(M)_\infty dP$ is a probability measure on Ω. By Theorem 1.8 for any local martingale X the process $\hat{X} = \langle X, M \rangle - X$ is a local martingale with respect to $d\hat{P}$ such that $\langle \hat{X} \rangle = \langle X \rangle$ under either probability measure. We shall show in Section 3.3 that if $M \in BMO$, then the Girsanov transformation $X \mapsto \hat{X}$ is an isomorphism of BMO onto $BMO(\hat{P})$. Here we shall only give the following remarkable result.

Corollary 2. 2.   *If there exists an innovation time T such that $P(T > 0) > 0$, then there exists a probability measure $\hat{P}$ equivalent to P such that $\hat{X} \notin H_1(\hat{P})$ for some bounded martingale X.*

Proof.   By Theorem 2.15 there is a bounded continuous martingale X which does not belong to $\bar{H}_\infty$. Obviously, $\langle X \rangle_\infty^{1/2}$ is not bounded. Since the dual of L_1 is L_∞, there is a random variable $W > 0$ such that $E[W] = 1$ and $E[W \langle X \rangle_\infty^{1/2}] = \infty$. Then, letting $d\hat{P} = W dP$, the conclusion follows immediately. $\qquad\square$

Remark 2.5.   We conjecture that if $(\mathcal{F}_t)$ is non-constant, then H_∞ as well as L_∞ is not closed in BMO. Now we give an example which supports this view. For that, consider the identity mapping S of $\mathbb{R}_+$ onto $\mathbb{R}_+$. Let μ be the probability measure on $\mathbb{R}_+$ defined by $\mu(S \in dx) = \sqrt{\frac{2}{\pi}} \exp(-\frac{x^2}{2}) dx$ and $\mathcal{G}_t$ be the μ-completion of the Borel field generated by $S \wedge t$. Then, S is a stopping time over $(\mathcal{G}_t)$. We next consider in the usual way a probability system $(\Omega, \mathcal{F}, P; (\mathcal{F}_t))$ by taking the product of the system $(\mathbb{R}_+, \mathcal{G}, \mu; (\mathcal{G}_t))$ with another system $(\Omega', \mathcal{F}', P'; (\mathcal{F}'_t))$ which carries a one dimensional Brownian motion $B = (B_t)$ starting at 0. Then the filtration $(\mathcal{F}_t)$ satisfies the usual conditions and S is also a stopping time over this filtration. Let M denote the process B stopped at S. It is a continuous martingale over $(\mathcal{F}_t)$ such that $\langle M \rangle_t = t \wedge S$. We first verify that $M \in BMO$.

Since $\{S > t\}$ is an $\mathcal{F}_t$-atom, we have

$$
\begin{aligned}
E[\langle M \rangle_\infty - \langle M \rangle_t | \mathcal{F}_t] &= E[S - t | \mathcal{F}_t] I_{\{t < S\}} \\
&\leq \left(\int_t^\infty \exp\left(-\frac{x^2}{2} \right) dx \right)^{-1} \int_t^\infty (x - t) \exp\left(-\frac{x^2}{2} \right) dx,
\end{aligned}
$$

which converges to 0 as $t \to \infty$. So there is a constant $C > 0$ such that $E[\langle M \rangle_\infty - \langle M \rangle_t | \mathcal{F}_t] \leq C$ for every t. This yields that $M \in BMO$. As a matter of course, we have $M \notin H_\infty$. Next, let $M^{(n)} = B^{n \wedge S}$, which belongs to the class H_∞. Since $\langle M^{(n)} - M \rangle_t = t \wedge S - t \wedge n \wedge S$, we find

$$
\begin{aligned}
&E[\langle M^{(n)} - M \rangle_\infty - \langle M^{(n)} - M \rangle_t | \mathcal{F}_t] \\
&\leq \left(\int_{t \vee n}^\infty \exp(-\frac{x^2}{2}) dx \right)^{-1} \int_{t \vee n}^\infty (x - t \vee n) \exp\left(-\frac{x^2}{2} \right) dx,
\end{aligned}
$$

so that $M^{(n)}$ converges in BMO to M as $n \to \infty$. Cosequently, $M \in \bar{H}_\infty \setminus H_\infty$.

Chapter 3

Exponential of BMO

3.1 The reverse Hölder inequality

We begin with a remark concerning the Muckenhoupt (A_p) condition. Let $0 \leq w(x) \in L^1_{loc}(R^n)$. In [5] R. R. Coifman and C. Fefferman proved that if w satisfies (A_p) for some $p > 1$, then the inequality

$$(3.1) \qquad \left(\frac{1}{|Q|} \int_Q w(x)^{1+\delta} dx \right)^{\frac{1}{1+\delta}} \leq C \left(\frac{1}{|Q|} \int_Q w(x) dx \right)$$

holds for all cubes Q, with constants $C, \delta > 0$ independent of Q. This is called the reverse Hölder inequality.

Now, let us give a probabilistic version of (3.1). For that, let M be a continuous local martingale, and consider the associated exponential martingale $\mathcal{E}(M)$. If $\mathcal{E}(M)$ is a uniformly integrable martingale, then

$$(3.2) \qquad \mathcal{E}(M)_T^p \leq E[\mathcal{E}(M)_\infty^p | \mathcal{F}_T]$$

for every p with $1 \leq p < \infty$, where T is a stopping time. This follows immediately from the conditional Hölder inequality.

Definition 3. 1. *Let $1 < p < \infty$. We say that $\mathcal{E}(M)$ satisfies (R_p) So we if the reverse Hölder inequality*

$$(3.3) \qquad E[\mathcal{E}(M)_\infty^p | \mathcal{F}_T] \leq C_p \mathcal{E}(M)_T^p$$

holds for every stopping time T, with a constant $C_p > 0$ depending only on p.

From the conditional Hölder inequality it follows at once that if $1 < p < r$, then (R_r) implies (R_p).

In this section we shall claim that if $M \in BMO$, then $\mathcal{E}(M)$ satisfies (R_p) for some $p > 1$ and that the converse statement is valid whenever $\mathcal{E}(M)$ is a uniformly integrable martingale. Let now $M \in BMO$. By Theorem 2.3 $\mathcal{E}(M)$ is a uniformly integrable martingale. Then, roughly speaking, $\mathcal{E}(M)$ satisfies a stronger reverse Hölder inequality as M gets near L_∞ in BMO. To see this, we set

$$(3.4) \qquad \Phi(x) = \left\{ 1 + \frac{1}{x^2} \log \frac{2x-1}{2(x-1)} \right\}^{\frac{1}{2}} - 1 \quad (1 < x < \infty),$$

which is clearly a continuous decreasing function such that $\Phi(1+0) = \infty$ and $\Phi(\infty) = 0$.

The reverse Hölder inequality for $\mathcal{E}(M)$ was first obtained by C. Doleans-Dade and P. A. Meyer ([11]). Recently, M. Emery ([15]) has given another proof of their result. The following is obtained by examining carefully the proof of Emery.

Theorem 3. 1.   Let $1 < p < \infty$. If $\|M\|_{BMO_2} < \Phi(p)$, then $\mathcal{E}(M)$ satisfies (R_p).

Proof.   We exclude the trivial case $\|M\|_{BMO_2} = 0$, and let us set

$$n(M) = 2\|M\|_{BMO_1} + \|M\|_{BMO_2}^2$$

for convenience' sake.

Suppose now $\|M\|_{BMO_2} < \Phi(p)$. Then we have

$$n(M) \leq (\|M\|_{BMO_2} + 1)^2 - 1 < \frac{1}{p^2} \log \frac{2p - 1}{2(p - 1)}$$

and so $0 < 2(p - 1)(2p - 1)^{-1} \exp\{p^2 n(M)\} < 1$. The main point in proving the theorem is to verify that

$$(3.5) \qquad E[\mathcal{E}(M)_\infty^p] \leq \frac{2}{1 - 2(p - 1)(2p - 1)^{-1} \exp\{p^2 n(M)\}}.$$

For simplicity, let $K_{p,M}$ denote the right hand side. For any stopping time T we have $n(M^T) \leq n(M)$ and so $K_{p,M^T} \leq K_{p,M}$. Therefore, in order to show (3.3), we may assume that $\mathcal{E}(M)$ is bounded. Next, let $\delta = \exp\{-pn(M)\}$, which is smaller that 1. A key to the proof of (3.5) is to use the following inequality :

$$(3.6) \qquad E[\mathcal{E}(M)_\infty : \mathcal{E}(M)_\infty > \lambda] \leq \frac{2p\lambda}{2p - 1} P(\mathcal{E}(M)_\infty > \delta\lambda) \quad (\lambda > 1).$$

In order to prove this inequality, let us define the stopping time $T = \inf\{t : \mathcal{E}(M)_t > \lambda\}$. Noticing $\log 1/\delta = pn(M)$, we find

$$\begin{aligned}
&P(\mathcal{E}(M)_\infty / \mathcal{E}(M)_T < \delta | \mathcal{F}_T) \\
&= P(1/\delta < \mathcal{E}(M)_T / \mathcal{E}(M)_\infty) \\
&= P\left(pn(M) < M_T - M_\infty + \frac{1}{2}(\langle M \rangle_\infty - \langle M \rangle_T) | \mathcal{F}_T\right) \\
&\leq \frac{1}{2pn(M)}\{2E[|M_\infty - M_T| + (\langle M \rangle_\infty - \langle M \rangle_T)|\mathcal{F}_T]\} \\
&\leq \frac{n(M)}{2pn(M)} = \frac{1}{2p},
\end{aligned}$$

so that

$$P(\mathcal{E}(M)_\infty / \mathcal{E}(M)_T \geq \delta | \mathcal{F}_T) \geq 1 - \frac{1}{2p}.$$

In addition to it, noticing $\mathcal{E}(M)_T = \lambda$ on $\{T < \infty\}$, we can obtain

$$P(\mathcal{E}(M)_\infty \geq \delta\lambda | \mathcal{F}_T) \geq \frac{2p - 1}{2p} I_{\{T < \infty\}}.$$

Therefore, it follows that

$$
\begin{aligned}
E[\mathcal{E}(M)_\infty : \mathcal{E}(M)_\infty > \lambda] &\leq E[\mathcal{E}(M)_\infty : T < \infty] \\
&\leq E[\mathcal{E}(M)_T : T < \infty] \\
&\leq \lambda P(T < \infty) \\
&\leq \frac{2p\lambda}{2p-1} P(\mathcal{E}(M)_\infty \geq \delta\lambda).
\end{aligned}
$$

Then, multiplying both sides of (3.6) by $(p-1)\lambda^{p-2}$ and integrating with respect to λ on the interval $[1, \infty[$, we find

$$
E[\mathcal{E}(M)_\infty^p - \mathcal{E}(M)_\infty : \mathcal{E}(M)_\infty > 1]
$$
$$
\leq \frac{2(p-1)}{2p-1} E[\{\delta^{-1}\mathcal{E}(M)_\infty\}^p - 1 : \mathcal{E}(M)_\infty > \delta],
$$

that is,

$$
\left\{ 1 - \frac{2(p-1)}{(2p-1)\delta^p} \right\} E[\mathcal{E}(M)_\infty^p : \mathcal{E}(M)_\infty > 1] \leq 1 + \frac{2(p-1)}{(2p-1)\delta^p}.
$$

Obviously, this yields (3.5).

Secondly, let T be any fixed stopping time. For an arbitrary element A of $\mathcal{F}_T$ such that $P(A) > 0$, we set

$$
dP' = 1_A \frac{dP}{P(A)}, \quad \mathcal{F}'_t = \mathcal{F}_{T+t}, \quad M'_t = M_{T+t} - M_T \quad (0 \leq t < \infty).
$$

Clearly, dP' is a probability measure and the process $M' = (M'_t, \mathcal{F}'_t)$ is a martingale with respect to dP'. Note that

$$
\mathcal{E}(M')_t = \mathcal{E}(M)_{T+t} / \mathcal{E}(M)_T.
$$

An elementary calculation shows that

$$
\|M'\|_{BMO_r(P')} \leq \|M\|_{BMO_r(P)}
$$

for every $r > 1$. Thus $\|M'\|_{BMO_2(P')} < \Phi(p)$. Then, repeating the same argument as above, we get

$$
E'[\mathcal{E}(M')_\infty^p] \leq K_{p,M'},
$$

where $E'[\ ]$ denotes expectation over Ω with respect to dP' and $K_{p,M'}$ is the constant corresponding to $K_{p,M}$ in (3.5). Namely, we have

$$
E[\{\mathcal{E}(M)_\infty / \mathcal{E}(M)_T\}^p : A] \leq K_{p,M'} P(A).
$$

However, since $\|M'\|_{BMO_r(P')} \leq \|M\|_{BMO_r(P)}$, we have $n(M') \leq n(M)$ and so $K_{p,M'} \leq K_{p,M}$. Thus the inequality

$$
E[\mathcal{E}(M)_\infty^p : A] \leq K_{p,M} \mathcal{E}(M)_T^p P(A)
$$

is valid for every $A \in \mathcal{F}_T$. This yields the reverse Hölder inequality (R_p) for $\mathcal{E}(M)$. Hence the theorem is established. □

Remark 3.1.   In [35] we studied this problem in a general setting and proved that, if M is a BMO-martingale satisfying $\triangle M \geq -1 + \delta$ for some δ with $0 < \delta \leq 1$, then $\mathcal{E}(M)$ is a uniformly integrable martingale and it satisfies the reverse Hölder inequality (R_p) for some $p > 1$.

It is an immediate consequence of Theorem 3.1 that if $M \in BMO$, then $\mathcal{E}(M) - 1 \in H_1$. Moreover, we obtain the following.

Theorem 3. 2.   *The mapping $\varphi : M \mapsto \mathcal{E}(M) - 1$ of BMO into H_1 is continuous.*

Proof.   Let $M \in BMO$ and let $n(M) = 2\|M\|_{BMO_1} + \|M\|_{BMO_2}^2$ as in the proof of Theorem 3.1. Next, let us choose $p > 1$ such that

$$\frac{2(p-1)}{2p-1} \exp\left(p^2 n(M)\right) < \frac{1}{2}.$$

Then from (3.5) it follows that $E[\mathcal{E}(M)_\infty^p] \leq 4$. To prove the theorem, assuming that $M^{(k)} \longrightarrow M$ in BMO as $k \to \infty$, it suffices to verify that $\varphi(M^{(k)}) \longrightarrow \varphi(M)$ in H_1. Since $\sup_k \|M^{(k)}\|_{BMO_2} < \infty$, there exists a number $p > 1$ such that

$$\sup_k \frac{2(p-1)}{2p-1} \exp\left(p^2 n\left(M^{(k)}\right)\right) \leq \frac{1}{2}.$$

Then

$$\sup_k E[\mathcal{E}(M^{(k)})_\infty^p] \leq 4,$$

from which it follows that $\{\mathcal{E}(M^{(k)})^*\}_{k=1,2,\cdots}$ is uniformly integrable. Furthermore, it is easy to see that $\mathcal{E}(M^{(k)})_\infty$ converges in probability to $\mathcal{E}(M)_\infty$ as $k \to \infty$. Thus $\mathcal{E}(M^{(k)})_\infty \longrightarrow \mathcal{E}(M)_\infty$ in L_r for $1 < r < p$. By using the theorem of B.Davis and the classical inequality of J.L.Doob we get

$$
\begin{aligned}
\left\|\varphi(M^{(k)}) - \varphi(M)\right\|_{H_1} &= \left\|\mathcal{E}(M^{(k)}) - \mathcal{E}(M)\right\|_{H_1} \\
&\leq 2E\left[\left\{\mathcal{E}\left(M^{(k)}\right) - \mathcal{E}(M)\right\}^*\right] \\
&\leq 2s\left\|\mathcal{E}\left(M^{(k)}\right)_\infty - \mathcal{E}(M)_\infty\right\|_r \longrightarrow 0 \quad (k \to \infty),
\end{aligned}
$$

where $r^{-1} + s^{-1} = 1$. This completes the proof.   □

As is stated in Remark 2.3, the generalization to the right continuous BMO-martingales is impossible.

Corollary 3. 1.   *If $M \in BMO$, then $x \mapsto \varphi(xM)$ is a continuous mapping of R^1 into H_1.*

It is remarkable that, even if $M \in H_p$ for all $p > 0$, $\mathcal{E}(M)$ does not necessarily belong to the class H_1.

Example 3.1.   Let $B = (B_t, \mathcal{F}_t)$ be a one dimensional Brownian motion starting at 0 and let T be the stopping time defined by

$$T = \inf\{t : B_t \leq t - 1\}.$$

Observe that

$$P(T \in dt) = \frac{1}{\sqrt{2\pi t^3}} \exp\left\{-\frac{(t-1)^2}{2t}\right\} dt.$$

Let now $M = B^T$. Clearly $M \notin BMO$. But, as $B_T = T - 1$, we have

$$\mathcal{E}(M)_\infty = \exp\left(\frac{1}{2}T - 1\right)$$

and so

$$E\left[\exp\left(\frac{1}{2}\langle M\rangle_\infty\right)\right] = E\left[\exp\left(\frac{1}{2}T\right)\right] \le e.$$

Thus $M \in H_p$ for all $p > 0$ and by Novikov's criterion $\mathcal{E}(M)$ is a uniformly integrable martingale. Next, let $d\hat{P} = \mathcal{E}(M)_\infty dP$, which is obviously a probability measure. By Girsanov's theorem the process $\hat{M} = \langle M\rangle - M$ is a local martingale with respect to $\hat{P}$ such that $\langle \hat{M}\rangle = \langle M\rangle$. Then we have

$$\begin{aligned}
\hat{E}[\langle \hat{M}\rangle_\infty] &= E[\mathcal{E}(M)_\infty \langle M\rangle_\infty] \\
&= e^{-1} E\left[\exp\left(\frac{1}{2}T\right) T\right] \\
&= e^{-1} \int_0^\infty \exp\left(\frac{1}{2}t\right) t \frac{1}{\sqrt{2\pi t^3}} \exp\left\{-\frac{(t-1)^2}{2t}\right\} dt \\
&= \int_0^\infty \frac{1}{\sqrt{2\pi t}} \exp\left(-\frac{1}{2t}\right) dt = \infty,
\end{aligned}$$

so $\hat{M} \notin H_2(\hat{P})$. Thus $\mathcal{E}(M) - 1 \notin H_1$ by Theorem 1.9.

Recall that the class BMO depends on the underlying probability measure, and so in case of necessity we denote by $BMO(\hat{P})$ the BMO class relative to $\hat{P}$.

Theorem 3. 3. *Assume that $\mathcal{E}(M)$ is a uniformly integrable martingale. Then $M \in BMO(P)$ if and only if $\hat{M} \in BMO(\hat{P})$.*

Proof. Suppose first that $M \in BMO(P)$. Then $\mathcal{E}(M)$ satisfies the reverse Hölder inequality (R_p) for some $p > 1$ by Theorem 3.1. We denote by $\mathcal{E}(\hat{M})$ the exponential local martingale corresponding to the local martingale $\hat{M}$ under the new probability measure $\hat{P}$. It is easy to see that $\mathcal{E}(\hat{M}) = 1/\mathcal{E}(M)$. Then we find

$$\begin{aligned}
\hat{E}\left[\left\{\frac{\mathcal{E}(\hat{M})_T}{\mathcal{E}(\hat{M})_\infty}\right\}^{\frac{1}{q-1}} \Big| \mathcal{F}_T\right] &= E\left[\frac{\mathcal{E}(M)_\infty}{\mathcal{E}(M)_T}\left\{\frac{\mathcal{E}(M)_\infty}{\mathcal{E}(M)_T}\right\}^{p-1} \Big| \mathcal{F}_T\right] \\
&= E\left[\left\{\frac{\mathcal{E}(M)_\infty}{\mathcal{E}(M)_T}\right\}^p \Big| \mathcal{F}_T\right] \le C_p,
\end{aligned}$$

where $p^{-1} + q^{-1} = 1$. In other words, $\mathcal{E}(\hat{M})$ satisfies the (A_q) condition with respect to $\hat{P}$. Thus $\hat{M} \in BMO(\hat{P})$ by Theorem 2.4. The validity of the converse statement is obvious. $\square$

Combining Theorem 3.3 with Theorem 3.1 gives the following.

Theorem 3. 4.   *Assume that $\mathcal{E}(M)$ is a uniformly integrable martingale. Then $M \in BMO$ if and only if $\mathcal{E}(M)$ satisfies the reverse Hölder inequality (R_p) for some $p > 1$.*

Proof.   Assume that $\mathcal{E}(M) \in (R_p)$ for $p > 1$. From the proof of Theorem 3.3 it follows that $\mathcal{E}(\hat{M})$ satisfies the (A_q) condition where q is the exponent conjugate to p. Then we have $\hat{M} \in BMO(\hat{P})$ by Theorem 2.4 and so $M \in BMO$ by Theorem 3.3. The converse is just the same as Theorem 3.1, and so we omit its proof.   □

Unless otherwise stated, we assume that $\mathcal{E}(M)$ is a uniformly integrable martingale. It should be noted that Theorem 3.4 does not hold without this condition. In the following we give such an example.

Example 3.2.   Consider a one dimensional Brownian motion $B = (B_t, \mathcal{F}_t)$ starting at 0. Note that $\mathcal{E}(B)_\infty = 0$ by (1.3). Therefore, it is clear that $\mathcal{E}(B)$ has all (R_p). However, the Brownian motion B does not belong to the class BMO.

3.2   Gehring's inequality

Gehring's inequality in the real analysis has a close relationship to the Muckenhoupt (A_p) condition, and its probabilistic version was given in [11] by C. Doléans-Dade and P. A. Meyer. First of all we shall show it.

Theorem 3. 5 (F.W.Gehring [20]).   *Let U be a positive random variable. If there are three constants $K > 0, \beta > 0, \varepsilon$ $(0 < \varepsilon < 1)$ such that*

$$(3.7) \qquad \int_{\{U > \lambda\}} U dP \le K \lambda^\varepsilon \int_{\{U > \beta\lambda\}} U^{1-\varepsilon} dP$$

for every $\lambda > 0$, then there are constants $r > 1$ and $C > 0$, depending only on K, β and ε, such that

$$(3.8) \qquad\qquad E[U^r] \le C E[U]^r.$$

Proof.   We may assume that $0 < \beta < 1$ and $E[U] = 1$. We first deal with the case where U is bounded. Multiplying both sides of the inequality (3.7) by $a\lambda^{a-1}$ $(a > 0)$ and integrating with respect to λ over the interval $[1, \infty[$, we find

$$\int_{\{U > 1\}} U dP \int_1^U a\lambda^{a-1} d\lambda \le K \int_{\{U > \beta\}} U^{1-\varepsilon} dP \int_1^{U/\beta} a\lambda^{a-1+\varepsilon} d\lambda.$$

The left hand side equals

$$\int_{\{U > 1\}} (U^{1+a} - U) dP,$$

and the right hand side is

$$\frac{aK}{a + \varepsilon} \int_{\{U > \beta\}} U^{1-\varepsilon} \left\{ \left(\frac{U}{\beta} \right)^{a+\varepsilon} - 1 \right\} dP \ \le\ K_a \int_{\{U > \beta\}} U^{1+a} dP$$

$$\le\ K_a \int_{\{U > 1\}} U^{1+a} dP + K_a,$$

where $K_a = \dfrac{aK}{(a+\varepsilon)\beta^{a+\varepsilon}}$. Namely, we get

(3.9) $\qquad E[U^{1+a} : U > 1] - E[U : U > 1] \leq K_a E[U^{1+a} : U > 1] + K_a.$

Now, let us choose $a > 0$ such that $K_a < 1$, and let $r = 1 + a$. The existence of such an a is guaranteed by the fact that $K_a \to 0$ as $a \to 0$. The inequality (3.9) yields

$$
\begin{aligned}
(1 - K_a)E[U^r : U > 1] &\leq E[U : U > 1] + K_a \\
&\leq E[U] + 1 \\
&\leq 2,
\end{aligned}
$$

from which it follows that

$$
E[U^r] \leq \frac{2}{1 - K_a} + 1.
$$

Since $E[U] = 1$ by the assumption, letting $C = \frac{2}{1-K_a} + 1$ we obtain (3.8).

Next, we deal with the case where U is unbounded. In this case there are two sequences $\{\alpha_m\}$ in R^1 and $\{\omega_m\}$ in Ω such that $0 < \alpha_m \uparrow \infty$ and $U(\omega_m) = \alpha_m$ for every m. For each m we set

$$
d\mu_m = \mathrm{I}_{\{U < \alpha_m\}} dP + \beta_m d\varepsilon_m,
$$

where $\beta_m = \frac{1}{\alpha_m} \int_{\{U \geq \alpha_m\}} U dP$ and ε_m denotes the Dirac measure at ω_m. Note that $\mu_m(\Omega) \geq 1$. Each μ_m is a finite measure such that $\mu_m(U > \alpha_m) = 0$, which implies that $U \in L_\infty(d\mu_m)$. Furthermore, we get

$$
\int_\Omega U d\mu_m = E[U]
$$

and

$$
\int_\Omega U^r d\mu_m \geq \int_{\{U < \alpha_m\}} U^r dP \longrightarrow E[U^r] \quad (m \to \infty).
$$

Thus it suffices to verify that for the same constants K, β, ε as in (3.7) the inequality

(3.10) $\qquad \displaystyle\int_{\{U > \lambda\}} U d\mu_m \leq K\lambda^\varepsilon \int_{\{U > \beta\lambda\}} U^{1-\varepsilon} d\mu_m$

is valid for every $\lambda > 0$.

If $\lambda \geq \alpha_m$, then the left hand side of (3.10) is obviously zero and so the inequality (3.10) follows at once. On the other hand, if $\lambda < \alpha_m$, then we have

$$
\begin{aligned}
\int_{\{U > \lambda\}} U d\mu_m &= \int_{\{\lambda < U < \alpha_m\}} U dP + \beta_m \alpha_m \\
&= \int_{\{\lambda < U\}} U dP
\end{aligned}
$$

and, noticing that $\beta_m \lambda < \lambda < \alpha_m$,

$$
\begin{aligned}
\int_{\{U > \beta\lambda\}} U^{1-\varepsilon} d\mu_m &= \int_{\{\beta\lambda < U < \alpha_m\}} U^{1-\varepsilon} dP + \beta_m \alpha_m^{1-\varepsilon} \\
&\geq \int_{\{\beta\lambda < U < \alpha_m\}} U^{1-\varepsilon} dP + \int_{\{U \geq \alpha_m\}} U^{1-\varepsilon} dP \\
&\geq \int_{\{U > \beta\lambda\}} U^{1-\varepsilon} dP.
\end{aligned}
$$

Combining these estimates with (3.7) shows (3.10). Thus the proof is complete. □

The next remarkable facts follows from the Gehring theorem.

Corollary 3. 2.   *Assume that $\mathcal{E}(M)$ is a uniformly integrable martingale, and let $1 < p < \infty$. If $\mathcal{E}(M)$ satisfies the reverse Hölder inequality (R_p), then it satisfies $(R_{p'})$ for some $p' > p$.*

Proof.   Let $U = \mathcal{E}(M)_\infty^p$, and we shall show that U satisfies (3.7). To see this, for each $\lambda > 0$ let

$$T_\lambda = \inf\{t : \mathcal{E}(M)_t^p > \lambda\},$$

which is clearly a stopping time. From this definition it follows immediately that $\mathcal{E}(M)_{T_\lambda}^p \le \lambda$ and

$$\{U > \lambda\} \subset \{T_\lambda < \infty\} \subset \{U > \beta\lambda\}$$

for every β with $0 < \beta < 1$. Let now $0 < \varepsilon < 1 - \frac{1}{p}$ (i.e., $p - \varepsilon p > 1$). Then by Jensen's inequality

$$\mathcal{E}(M)_{T_\lambda}^{p-p\varepsilon} \le E[\mathcal{E}(M)_\infty^{p-p\varepsilon}|\mathcal{F}_{T_\lambda}] = E[U^{1-\varepsilon}|\mathcal{F}_{T_\lambda}].$$

If $\mathcal{E}(M) \in (R_p)$, then

$$
\begin{aligned}
\int_{\{U>\lambda\}} U\,dP &\le \int_{\{T_\lambda<\infty\}} E[\mathcal{E}(M)_\infty^p|\mathcal{F}_{T_\lambda}]dP \\
&\le K_p \int_{\{T_\lambda<\infty\}} \mathcal{E}(M)_{T_\lambda}^p\,dP \\
&\le K_p\lambda^\varepsilon \int_{\{T_\lambda<\infty\}} \mathcal{E}(M)_{T_\lambda}^{p-p\varepsilon}\,dP \\
&\le K_p\lambda^\varepsilon \int_{\{T_\lambda<\infty\}} U^{1-\varepsilon}\,dP \\
&\le K_p\lambda^\varepsilon \int_{\{U>\beta\lambda\}} U^{1-\varepsilon}\,dP.
\end{aligned}
$$

So, from Theorem 3.5 it follows that for some $r > 1$

$$E[U^r] \le C_r E[U]^r,$$

that is,

$$E[\mathcal{E}(M)_\infty^{pr}] \le C_r E[\mathcal{E}(M)_\infty^p]^r.$$

By the usual manner we can easily obtain the conditional form of this inequality as follows :

$$E[\{\mathcal{E}(M)_\infty/\mathcal{E}(M)_T\}^{pr}|\mathcal{F}_T] \le C_r E[\{\mathcal{E}(M)_\infty/\mathcal{E}(M)_T\}^p|\mathcal{F}_T]^r$$

where T is an arbitrary stopping time. Then, since $\mathcal{E}(M)$ satisfies (R_p), we have

$$E[\mathcal{E}(M)_\infty^{pr}|\mathcal{F}_T] \le C_{p,r}\mathcal{E}(M)_T^{pr},$$

which completes the proof.   □

It should be noted that the following result is in sympathy with Corollary 3.2.

Corollary 3. 3. *Let $1 < p < \infty$. If $\mathcal{E}(M)$ satisfies (A_p), then it satisfies $(A_{p-\varepsilon})$ for some ε with $0 < \varepsilon < p - 1$.*

Proof. Assume that $\mathcal{E}(M)$ satisfies (A_p). Then we have $M \in BMO$ by Theorem 2.4 and so, according to Theorem 2.3, $\mathcal{E}(M)$ is a uniformly integrable martingale. Let now $d\hat{P} = \mathcal{E}(M)_\infty dP$. By Theorem 1.8 the process $\hat{M} = \langle M \rangle - M$ is a martingale with respect to $d\hat{P}$ such that $\langle \hat{M} \rangle = \langle M \rangle$ under either probability. Recall that the associated exponential martingale $\mathcal{E}(\hat{M})$ is equal to $\mathcal{E}(M)^{-1}$. Let q be the exponent conjugate to p. Then it follows from the definition of the conditional expectation $\hat{E}[\ |\mathcal{F}_T]$ that

$$\hat{E}[\{\mathcal{E}(\hat{M})_\infty/\mathcal{E}(\hat{M})_T\}^q|\mathcal{F}_T] = E\left[\{\mathcal{E}(M)_T/\mathcal{E}(M)_\infty\}^{\frac{1}{p-1}}\Big|\mathcal{F}_T\right],$$

which is bounded by some constant C_p depending only on p. This implies that $\mathcal{E}(\hat{M})$ satisfies the reverse Hölder inequality (R_q) with respect to $\hat{P}$. Then $\mathcal{E}(\hat{M})$ satisfies $(R_{q'})$ for some $q' > q$ by Corollary 3.2. Let p' be the exponent conjugate to q'. As $p' < p$, replacing p and q with p' and q' respectively in the above inequality shows that $\mathcal{E}(M)$ satisfies the $(A_{p'})$ condition. Thus the proof is complete. □

Remark 3.2. Let M be a right continuous local martingale such that $-1 + \delta \leq \Delta M \leq C$ for some $C > 0$ and $\delta(0 < \delta \leq 1)$. In [11] C. Doléans-Meyer and P.A. Meyer proved that $\mathcal{E}(M)$ satisfies the same property as Corollary 3.3.

Let now $1 < p < \infty$. We say for convenience that $\mathcal{E}(M)$ satisfies (B_p) if for every stopping time T

$$(3.11) \qquad \mathcal{E}(M)_T^{1/p} \leq K_p E[\mathcal{E}(M)_\infty^{1/p}|\mathcal{F}_T]$$

where K_p is a positive constant depending only on p. This is in fact no other than the condition $(b_{1/p}^-)$ which is stated in [11]. If $1 < p < p'$, then $(B_{p'})$ implies (B_p), which follows from the Jensen inequality. Note that if $\mathcal{E}(M)$ is a uniformly integrable martingale, then the inequality (3.11) is clearly valid for $0 < p \leq 1$ with $K_p = 1$.

Corollary 3. 4. *If $M \in BMO$, then $\mathcal{E}(M)$ satisfies (B_p) for all $p > 1$. More precisely, the inequality*

$$(3.12) \qquad \mathcal{E}(M)_T^{1/p} \leq \exp\left(\frac{1}{2p}\|M\|_{BMO_2}^2\right) E[\mathcal{E}(M)_\infty^{1/p}|\mathcal{F}_T]$$

holds for every stopping time T.
  Conversely, if $\mathcal{E}(M)$ satisfies (B_p) for some $p > 1$, then $M \in BMO$.

Proof. Firstly, let $M \in BMO$. Then Jensen's inequality shows that for every $p > 1$ and every stopping time T

$$\begin{aligned}
E[\mathcal{E}(M)_\infty^{1/p}|\mathcal{F}_T] &= E[\{\mathcal{E}(M)_\infty/\mathcal{E}(M)_T\}^{1/p}|\mathcal{F}_T]\mathcal{E}(M)_T^{1/p}\\
&\geq \exp\left(-\frac{1}{2p}E[\langle M \rangle_\infty - \langle M \rangle_T|\mathcal{F}_T]\right)\mathcal{E}(M)_T^{1/p}\\
&\geq \exp\left(-\frac{1}{2p}\|M\|_{BMO_2}^2\right)\mathcal{E}(M)_T^{1/p},
\end{aligned}$$

so (3.12) holds.

Secondly, we shall verify the converse. For that, let us assume that $\mathcal{E}(M)$ satisfies (B_p) for some $p > 1$ and let $p^{-1} + q^{-1} = 1$. The uniform integrability of $\mathcal{E}(M)$ is not yet confirmed, but, applying Jensen's inequality shows

$$\mathcal{E}(M)_T \leq KE[\mathcal{E}(M)_\infty | \mathcal{F}_T],$$

from which it follows at once that $\mathcal{E}(M)$ is a uniformly integrable martingale. Therefore, according to Theorem 3.4, it is enough to verify that $\mathcal{E}(M)$ satisfies (R_r) for some $r > 1$. We show it by using the same idea as in the proof of Corollary 3.2. For each $\lambda > 0$, let T_λ be the stopping time defined by

$$T_\lambda = \inf\{t : \mathcal{E}(M)_t > \lambda\}.$$

Then $\mathcal{E}(M)_{T_\lambda} \leq \lambda$ clearly and for every β with $0 < \beta < 1$

$$\{\mathcal{E}(M)_\infty > \lambda\} \subset \{T_\lambda < \infty\} \subset \{\mathcal{E}(M)_\infty > \beta\lambda\}.$$

Since $\mathcal{E}(M)$ satisfies (3.12) by the assumption, we have

$$
\begin{aligned}
\int_{\{\mathcal{E}(M)_\infty > \lambda\}} \mathcal{E}(M)_\infty dP &\leq \int_{\{T_\lambda < \infty\}} \mathcal{E}(M)_{T_\lambda} dP \\
&\leq \lambda^{1/q} \int_{\{T_\lambda < \infty\}} \mathcal{E}(M)_{T_\lambda}^{1/p} dP \\
&\leq K_p \lambda^{1/q} \int_{\{\mathcal{E}(M)_\infty > \beta\lambda\}} \mathcal{E}(M)_\infty^{1/p} dP \\
&\leq K_p \lambda^{1/q} \int_{\{\mathcal{E}(M)_\infty > \beta\lambda\}} \mathcal{E}(M)_\infty^{1-1/q} dP.
\end{aligned}
$$

Thus it follows from Theorem 3.5 that $\mathcal{E}(M)$ satisfies (R_r) for some $r > 1$. This completes the proof.   □

3.3   Transformation of BMO by a change of law

Let us assume that $\mathcal{E}(M)$ is a uniformly integrable martingale and let $d\hat{P} = \mathcal{E}(M)_\infty dP$ as before. Since the classes BMO and H_p depend on the underlying probability measure, we denote them by such as $BMO(P)$ and $H_p(P)$ in case of necessity. We already know that this change of law yields the Girsanov transformation $\phi : \mathcal{L}(P) \ni X \mapsto \hat{X} = \langle X, M \rangle - X \in \mathcal{L}(\hat{P})$ where $\mathcal{L}$ denotes the class of all continuous local martingales. As remarked in Section 2.6 of Chapter 2, both BMO and H_p are not invariant under the Girsanov transformation. In this section we shall show that if $M \in BMO(P)$, then this trouble can be removed.

Theorem 3. 6.   If $M \in BMO(P)$, then $\phi : X \mapsto \hat{X}$ is an isomorphism of $BMO(P)$ onto $BMO(\hat{P})$.

Proof.   Assume that $M \in BMO(P)$. Then $\hat{M} \in BMO(\hat{P})$ by Theorem 3.3, so that for some $p > 1$ the associated exponential martingale $\mathcal{E}(\hat{M})$ satisfies (A_p) by Theorem 2.4 : that is, for any stopping time T we have

$$(3.13) \qquad \hat{E}\left[\left\{\mathcal{E}(\hat{M})_T / \mathcal{E}(\hat{M})_\infty\right\}^{\frac{1}{p-1}} \Big| \mathcal{F}_T\right] \leq C_p.$$

Our claim is that the inequality

$$\|\phi(X)\|_{BMO_2(\hat{P})} \leq C \|X\|_{BMO_2(P)} \tag{3.14}$$

is valid for all $X \in BMO(P)$, with a constant $C > 0$ independent of X. Since $\|X\|_{BMO_2(P)} = 0$ implies $\phi(X) = 0$, we may assume that $\|X\|_{BMO_2(P)} > 0$. Let now $\alpha = (2p\|X\|_{BMO_2(P)}^2)^{-1}$. Then from Theorem 2.2 we get

$$E[\exp\{\alpha p(\langle X \rangle_\infty - \langle X \rangle_T)\}|\mathcal{F}_T] \leq \frac{1}{1 - \alpha p \|X\|_{BMO_2(P)}^2} = 2. \tag{3.15}$$

Noticing the fact that $\langle \phi(X) \rangle = \langle X \rangle$ under either probability measure and applying the Hölder inequality with exponents p and $\frac{p}{p-1}$ we have

$$\hat{E}[\langle \phi(X) \rangle_\infty - \langle \phi(X) \rangle_T | \mathcal{F}_T]$$

$$\leq \frac{1}{\alpha} \hat{E}\left[\left\{ \frac{\mathcal{E}(\hat{M})_T}{\mathcal{E}(\hat{M})_\infty} \right\}^{1/p} \left\{ \frac{\mathcal{E}(\hat{M})_\infty}{\mathcal{E}(\hat{M})_T} \right\}^{1/p} \exp\{\alpha(\langle X \rangle_\infty - \langle X \rangle_T)\} | \mathcal{F}_T \right]$$

$$\leq 2p\|X\|_{BMO_2(P)}^2 \hat{E}\left[\left\{ \frac{\mathcal{E}(\hat{M})_T}{\mathcal{E}(\hat{M})_\infty} \right\}^{\frac{1}{p-1}} \Big| \mathcal{F}_T \right]^{(p-1)/p} E[\exp\{\alpha p(\langle X \rangle_\infty - \langle X \rangle_T)\}|\mathcal{F}_T]^{1/p}.$$

By (3.13) the first conditional expectation in the last expression is bounded and, according to (3.15), the second one is bounded by 2. Thus (3.14) holds. From the closed graph theorem it follows that the similar inequality holds for the inverse mapping ϕ^{-1}. Hence $BMO(P)$ and $BMO(\hat{P})$ are isomorphic under the mapping ϕ.   □

I conjecture that the converse of Theorem 3.6 must be true. Note that, without the condition "$M \in BMO$", $\phi(X)$ does not always belong to the class $H_1(P)$ even if X is bounded. See Corollary 2.2 in Chapter 2.

Now, let $1 \leq p < \infty$ and let us define the mapping $\Phi_p : \mathcal{L}(P) \mapsto \mathcal{L}(\hat{P})$ by $\Phi_p(X) = \mathcal{E}(M)^{-1/p} \circ \phi(X)$, that is, for every $t \geq 0$

$$\Phi_p(X)_t = \int_0^t \mathcal{E}(M)_s^{-1/p} d\phi(X)_s, \quad (X \in \mathcal{L}(P)). \tag{3.16}$$

From the definition of ϕ it follows at once that $\Phi_p(X) = \phi(\mathcal{E}(M)^{-1/p} \circ X)$, so that Φ_p is also a linear bijection of $\mathcal{L}(P)$ onto $\mathcal{L}(\hat{P})$.

Theorem 3. 7. Let $1 \leq p < \infty$. If $M \in BMO(P)$, then $H_p(P)$ and $H_p(\hat{P})$ are isomorphic under the mapping Φ_p.
In particular, if $1 < p < \infty$, then

$$\exp\left(-\frac{1}{2q}\|\phi(M)\|_{BMO_2(\hat{P})}^2 \right) \leq \|\Phi_p\| \leq \exp\left(\frac{1}{2q}\|M\|_{BMO_2(P)}^2 \right) \tag{3.17}$$

where $p^{-1} + q^{-1} = 1$ and $\|\Phi_p\|$ denotes the norm of Φ_p as an operator from $H_p(P)$ to $H_p(\hat{P})$.

Proof. We first deal with the case $1 < p < \infty$. Let $p^{-1} + q^{-1} = 1$. If $M \in BMO(P)$, then it follows from Corollary 3.4 that $\mathcal{E}(M)$ is a uniformly integrable martingale which satisfies the inequality

$$(3.18) \qquad \mathcal{E}(M)_T^{1/q} \leq A_q E\left[\mathcal{E}(M)_\infty^{1/q} | \mathcal{F}_T\right]$$

where $A_q = \exp(\frac{1}{2q} \|M\|_{BMO_2(P)}^2)$. Let now $X \in H_p(P)$ and $Y' \in H_q(\hat{P})$.
By using the stopping time argument we may assume that the increasing process $\{\int_0^t \mathcal{E}(M)_s^{-1/p} |d\langle \hat{X}, Y'\rangle_s|\}$ is integrable with respect to $d\hat{P}$. Then, combining (3.18) with the Hölder inequality, we find

$$
\begin{aligned}
\hat{E}[|\langle \Phi_p(X), Y'\rangle_\infty|] &\leq E\left[\mathcal{E}(M)_\infty \int_0^\infty \mathcal{E}(M)_t^{-1/p} |d\langle \hat{X}, Y'\rangle_t|\right] \\
&\leq E\left[\int_0^\infty \mathcal{E}(M)_t^{1/q} |d\langle \hat{X}, Y'\rangle_t|\right] \\
&\leq A_q E\left[\int_0^\infty E[\mathcal{E}(M)_\infty^{1/q} | \mathcal{F}_t] |d\langle \hat{X}, Y'\rangle_t|\right] \\
&\leq A_q E\left[\mathcal{E}(M)_\infty^{1/q} \int_0^\infty |d\langle \hat{X}, Y'\rangle_t|\right] \\
&\leq A_q E\left[\mathcal{E}(M)_\infty^{1/q} \langle \hat{X}\rangle_\infty^{1/2} \langle Y'\rangle_\infty^{1/2}\right] \\
&\leq A_q E\left[\langle \hat{X}\rangle_\infty^{p/2}\right]^{1/p} \|Y'\|_{H_q(\hat{P})} \\
&\leq A_q \|X\|_{H_p(P)} \|Y'\|_{H_q(\hat{P})}.
\end{aligned}
$$

Thus the usual duality argumemt yields

$$\|\Phi_p(X)\|_{H_p(\hat{P})} \leq A_q \|X\|_{H_p(P)},$$

from which the right-hand side of (3.17) follows at once.
On the other hand, $\hat{M} = \phi(M)$ belongs to the class $BMO(\hat{P})$ by Theorem 3.3, so that the linear mapping $\hat{\Phi}_p : H_p(\hat{P}) \mapsto H_p(P)$ defined by

$$\hat{\Phi}_p(X')_t = \int_0^t \mathcal{E}(\hat{M})_s^{-1/p} d\phi^{-1}(X')_s \quad (t \geq 0, X' \in H_p(\hat{P}))$$

is continuous. More precisely, in the same manner as the case for Φ_p we can verify that for every $X' \in H_p(\hat{P})$

$$\|\hat{\Phi}_p(X')\|_{H_p(P)} \leq \hat{A}_q \|X'\|_{H_p(\hat{P})}$$

where $\hat{A}_q = \exp(\frac{1}{2q} \|\hat{M}\|_{BMO_2(\hat{P})}^2)$. Observe that $\hat{\Phi}_p$ is the inverse of Φ_p. Consequently, the mapping Φ_p is an isomorphism of $H_p(P)$ onto $H_p(\hat{P})$. Then for every $X \in H_p(P)$ we have

$$
\begin{aligned}
\|X\|_{H_p(P)} &= \|\hat{\Phi}_p(\Phi_p(X))\|_{H_p(P)} \\
&\leq \hat{A}_q \|\Phi_p(X)\|_{H_p(\hat{P})} \\
&\leq \hat{A}_q \|\Phi_p\| \|X\|_{H_p(\hat{P})},
\end{aligned}
$$

from which the left-hand side of (3.17) follows.

Secondly, we shall establish the isomorphism between $H_1(P)$ and $H_1(\hat{P})$. Recall that for all $Y \in BMO(P)$

$$(3.19) \qquad c_1 \|Y\|_{BMO_2(P)} \leq \|\phi(Y)\|_{BMO_2(\hat{P})} \leq c_2 \|Y\|_{BMO_2(P)}$$

with the choice of these positive constants c_1 and c_2 depending only on M. This is just a consequence of Theorem 3.6, and observe that if $Y' \in BMO(\hat{P})$, then $Y' = \phi(Y)$ for some $Y \in BMO(P)$. Combining Fefferman's inequality with the left side of (3.19) yields that if $X \in H_1(P)$ and $Y' \in BMO(\hat{P})$, then

$$
\begin{aligned}
\hat{E}[\langle \Phi_1(X), Y' \rangle_\infty] &\leq \hat{E}[\int_0^\infty |d\langle \Phi_1(X), Y' \rangle_t|] \\
&\leq E\left[\mathcal{E}(M)_\infty \int_0^\infty \mathcal{E}(M)_t^{-1} |d\langle \phi(X), \phi(Y) \rangle_t|\right] \\
&\leq E\left[\int_0^\infty |d\langle X, Y \rangle_t|\right] \\
&\leq \sqrt{2}\|X\|_{H_1(P)}\|Y\|_{BMO_2(P)} \\
&\leq \sqrt{2}c_1^{-1}\|X\|_{H_1(P)}\|Y'\|_{BMO_2(\hat{P})}.
\end{aligned}
$$

This implies that

$$\|\Phi_1(X)\|_{H_1(\hat{P})} \leq \sqrt{2}c_1^{-1}\|X\|_{H_1(P)}.$$

Let now

$$\hat{\Phi}_1(X')_t = \int_0^t \mathcal{E}(\hat{M})_s^{-1} d\phi^{-1}(X')_s$$

where X' is a local martingale with respect to $\hat{P}$. Note that the mapping $\hat{\Phi}_1 : \mathcal{L}(\hat{P}) \mapsto \mathcal{L}(P)$ is nothing but the inverse of Φ_1. Repeating the similar argument for $X' \in H_1(\hat{P})$ and $Y \in BMO(P)$ we can verify that

$$E\left[\int_0^\infty |d\langle \hat{\Phi}_1(X'), Y \rangle_t|\right] \leq \sqrt{2}c_2\|X'\|_{H_1(\hat{P})}\|Y\|_{BMO_2(P)}.$$

Then we have

$$\|\Phi_1^{-1}(X')\|_{H_1(\hat{P})} \leq \sqrt{2}c_2\|X'\|_{H_1(\hat{P})}$$

for all $X' \in H_1(\hat{P})$. Hence, $H_1(P)$ and $H_1(\hat{P})$ are isomorphic under the mapping Φ_1. This completes the proof. □

It should be noted that the assumption "$M \in BMO(P)$" can not be omitted for the validity of the theorem. In the following we give such an example.

Example 3.3. Let $(\Omega, \mathcal{F}, P), B = (B_t, \mathcal{F}_t)$ and S be as in Example 2.1, except the distribution of S. In order to make sure we explain them repeatedly. Let $\mathcal{G}^\circ$ be the class of all topological Borel sets in R_+ and let S be the identity mapping of R_+ onto R_+. We now define a probability measure $d\mu$ on R_+ such that

$$d\mu = I_{[1,\infty)}(t)t^{-2}dt.$$

Let $\mathcal{G}$ be the completion of $\mathcal{G}^\circ$ with respect to $d\mu$ and let $\mathcal{G}_t$ be the completion of the Borel field generated by $S \wedge t$. Note that S is clearly a stopping time over $(\mathcal{G}_t)$. Let now $B = (B_t, \mathcal{F}'_t)$ be a one dimensional Brownian motion starting at 0 defined on a probability space $(\Omega', \mathcal{F}', P')$, and let us construct in the usual way a probability system $(\Omega, \mathcal{F}, P; (\mathcal{F}_t))$ by taking the product of the system $(R_+, \mathcal{G}, d\mu; (\mathcal{G}_t))$ with $(\Omega', \mathcal{F}', P'; (\mathcal{F}'_t))$. Then S is also a stopping time over $(\mathcal{F}_t)$, so that $M = B^S$ is a continuous martingale over this filtration. It is not difficult to see that $E[\mathcal{E}(M)_\infty] = 1$ and $E[\langle M \rangle_\infty^{1/2}] = \int_1^\infty t^{-3/2} dt < \infty$. The latter fact means that $M \in H_1(P)$. However, $\Phi_1(M)$ does not belong to $H_1(\hat{P})$. In fact, since $E[\langle M \rangle_\infty] = \int_1^\infty t^{-1} dt = \infty$, M does not belong to $H_2(P)$. Then $\mathcal{E}(\hat{M}) - 1 \notin H_1(\hat{P})$ by Theorem 1.9, so that $\Phi_1(M) = \mathcal{E}(M)^{-1} \circ \phi(M) \notin H_1(\hat{P})$, because $\mathcal{E}(M)^{-1} \circ \phi(M) = \mathcal{E}(\hat{M}) \circ \hat{M} = \mathcal{E}(\hat{M}) - 1$.

Remark 3.3.   A generalization of Theorem 3.7 to non-continuous martingales is obrained in [37] as follows : Let $1 \le p < \infty$. If $M \in BMO$ and $\triangle M \ge -1 + \delta$ for some δ with $0 < \delta \le 1$, then the mapping $X \mapsto \mathcal{E}(M)_-^{-1/p} \circ \phi(X)$ is an isomorphism of H_p onto $\hat{H}_p$ where ϕ denotes the Girsanov transformation. Particularly, the spaces BMO and $BMO(\hat{P})$ are isomorphic under ϕ.

3.4   A characterization of the BMO-closure of L_∞

In Section 2.5 we have obtained a characterization of $\bar{L}_\infty$ which is a probabilistic version of the result due to J. Garnett and P. Jones. The aim of this section is to give a very interesting characterization of $\bar{L}_\infty$ in the framework of exponential martingales as follows.

Theorem 3. 8.   *Let $M \in BMO$. Then $M \in \bar{L}_\infty$ if and only if $\mathcal{E}(\lambda M)$ satisfies all (R_p) for every real number λ.*

Proof.   We first show the "only if" part. For that, suppose $M \in \bar{L}_\infty$. Then $a(M) = \infty$ by Theorem 2.8. On the other hand, for every λ and every stopping time T

$$\frac{\mathcal{E}(\lambda M)_\infty}{\mathcal{E}(\lambda M)_T} \le \exp(|\lambda||M_\infty - M_T|)$$

by the definition of $\mathcal{E}(\lambda M)$. Therefore, recalling the definition of $a(M)$, we find

$$
\begin{aligned}
E[\mathcal{E}(\lambda M)_\infty^p | \mathcal{F}_T] &= E\left[\left\{\frac{\mathcal{E}(\lambda M)_\infty}{\mathcal{E}(\lambda M)_T}\right\}^p \Big| \mathcal{F}_T\right] \mathcal{E}(\lambda M)_T^p \\
&\le E[\exp(|\lambda| p |M_\infty - M_T|) | \mathcal{F}_T] \mathcal{E}(\lambda M)_T^p \\
&\le C_{\lambda, p} \mathcal{E}(\lambda M)_T^p
\end{aligned}
$$

for every λ and every $p > 1$, with a constant $C_{\lambda, p} > 0$ independent of T.

We are now going to prove the "if" part. By the John-Nirenberg theorem there exists some positive number α_0 such that for every stopping time T

$$E\left[\exp\left\{\frac{\alpha_0}{2}(\langle M \rangle_\infty - \langle M \rangle_T)\right\} \Big| \mathcal{F}_T\right] \le C_0,$$

where C_0 is a positive constant independent of T. Next, let $\lambda > 0$ and $0 < \alpha < \min\{2\lambda, \alpha_0/(2\lambda)\}$. If we set $p = 2\lambda/\alpha$, then $1 < p < \alpha_0/\alpha^2$ and so

$$
\exp\left\{2\lambda(M_\infty - M_T) - \frac{\alpha_0}{2}(\langle M\rangle_\infty - \langle M\rangle_T)\right\}
$$
$$
= \exp\left\{\frac{2\lambda}{\alpha}(\alpha M_\infty - \alpha M_T) - \frac{1}{2}\frac{\alpha_0}{\alpha^2}(\langle \alpha M\rangle_\infty - \langle \alpha M\rangle_T)\right\}
$$
$$
\leq \left\{\frac{\mathcal{E}(\alpha M)_\infty}{\mathcal{E}(\alpha M)_T}\right\}^p.
$$

Combining this with the Schwarz inequality gives

$$
E[\exp\{\lambda(M_\infty - M_T)\}|\mathcal{F}_T]
$$
$$
= E\left[\exp\left\{\lambda(M_\infty - M_T) - \frac{\alpha_0}{4}(\langle M\rangle_\infty - \langle M\rangle_T)\right\}\right.
$$
$$
\left. \cdot \exp\left\{\frac{\alpha_0}{4}(\langle M\rangle_\infty - \langle M\rangle_T)\right\}\Big|\mathcal{F}_T\right]
$$
$$
\leq E[\{\mathcal{E}(\alpha M)_\infty/\mathcal{E}(\alpha M)_T\}^p|\mathcal{F}_T]^{1/2} E\left[\exp\left\{\frac{\alpha_0}{2}(\langle M\rangle_\infty - \langle M\rangle_T)\right\}\Big|\mathcal{F}_T\right]^{1/2}.
$$

Since $\mathcal{E}(\alpha M)$ satisfies (R_p) by the assumption, the first conditional expectation in the last expression is dominated by some constant. Furthermore, the second term is smaller than $C_0^{1/2}$ as previosly stated. Therefore, we have

$$
E[\exp\{\lambda(M_\infty - M_T)\}|\mathcal{F}_T] \leq C_\lambda
$$

for every stopping time T, with a constant $C_\lambda > 0$ depending only on λ. The same argument works if M is replaced by $-M$, so that for every $\lambda > 0$

$$
E[\exp(\lambda|M_\infty - M_T|)|\mathcal{F}_T] \leq C_\lambda,
$$

where T is an arbitrary stopping time. This implies that $a(M) = \infty$. Then it follows at once from Theorem 2.8 that $M \in \bar{L}_\infty$. Thus the theorem is established. $\qquad \square$

From the above proof it follows that if $\mathcal{E}(\lambda M)$ satisfies all (R_p) only for λ sufficiently small in absolute value, then M belongs to $\bar{L}_\infty$. So it is natural to ask the question whether or not $\mathcal{E}(\alpha M)$ has (R_s) for $\alpha < 0$ and $s > 1$ if $\mathcal{E}(\lambda M)$ has all (R_p) for every $\lambda > 0$. Taking this question simple and easy, it seems to be true. But, surprisingly enough, the answer in general is no. We give below such an example.

Example 3.4. Let $B = (B_t, \mathcal{F}_t)$ be a one dimensional Brownian motion with $B_0 = 0$ defined on a probability space $(\Omega, \mathcal{F}, Q)$ and let $\tau = \inf\{t : |B_t| = 1\}$. Then B^τ is cleary a bounded martingale, so that $dP = \exp(B_\tau - \frac{1}{2}\tau)dQ$ is also a probability measure on Ω. Consider now the process $M = 2B^\tau - 2\langle B^\tau\rangle$, which is a BMO-martingale with respect to dP by Theorem 3.6. Noticing $\langle M\rangle_t = 4(t \wedge \tau)$, we find that

$$
E[\{\mathcal{E}(\lambda M)_\infty/\mathcal{E}(\lambda M)_T\}^p|\mathcal{F}_T]
$$
$$
= E_Q\left[\exp\left\{(B_\tau - B_{\tau \wedge T} - \frac{1}{2}(\tau - \tau \wedge T)\right\}\right.
$$

$$\cdot \exp\left\{ p\lambda(M_\infty - M_T) - \frac{1}{2}p\lambda^2(\langle M\rangle_\infty - \langle M\rangle_T)\right\}\Big| \mathcal{F}_T\right]$$

$$= E_Q\left[\exp\{(1 + 2p\lambda)(B_\tau - B_{\tau\wedge T})\}\right.$$

$$\cdot \exp\left\{-\frac{1}{2}\left(4p\lambda^2 + 4p\lambda + 1\right)(\tau - \tau\wedge T)\right\}\Big| \mathcal{F}_T\right],$$

where $E_Q[\ ]$ denotes expectation with respect to dQ.
Thus, if $4p\lambda^2 + 4p\lambda + 1 \geq 0$ (that is, $|\lambda + \frac{1}{2}| \geq \frac{1}{2\sqrt{q}}$ where $\frac{1}{p} + \frac{1}{q} = 1$), then we have

$$E[\{\mathcal{E}(\lambda M)_\infty / \mathcal{E}(\lambda M)_T\}^p | \mathcal{F}_T] \leq \exp\{2(1 + 2p|\lambda|)\}.$$

This means that $\mathcal{E}(\lambda M)$ has all (R_p) whenever $\lambda > 0$ or $\lambda \leq -1$. Particularly, both $\mathcal{E}(M)$ and $\mathcal{E}(-M)$ have all (R_p).
On the other hand, if $-1 < \lambda < 0$, then $\mathcal{E}(\lambda M)$ does not satisfy (R_p) for $p \geq \frac{1 + \pi^2/4}{1 - (2\lambda + 1)^2}$. To verify it, recall that $E_Q[\exp(\alpha\tau)] = \infty$ for $\alpha \geq \frac{\pi^2}{8}$ by Lemma 1.3. Since $-(4p\lambda^2 + 4p\lambda + 1) \geq \frac{\pi^2}{4}$ for such λ and p, we have

$$E[\mathcal{E}(\lambda M)_\infty^p] \geq \exp\{-(1 + 2p)\}E_Q\left[\exp\left\{-\frac{1}{2}(4p\lambda^2 + 4p\lambda + 1)\tau\right\}\right] = \infty,$$

which implies that $\mathcal{E}(\lambda M)$ does not satisfy the (R_p) condition. Then it naturally follows from Theorem 3.8 that $M \notin \bar{L}_\infty$.
We estimate in parentheses the distance between M and L_∞. If $\lambda \geq \frac{1}{4} + \frac{\pi^2}{16}$, then $2\lambda - \frac{1}{2} \geq \frac{\pi^2}{8}$ clearly and so we find

$$\begin{aligned} E[\exp(-\lambda M_\infty)] &= E_Q\left[\exp\left(B_\tau - \frac{\tau}{2}\right)\exp(-2\lambda B_\tau + 2\lambda\tau)\right] \\ &= E_Q\left[\exp\{(1 - 2\lambda)B_\tau\}\exp\left\{\left(2\lambda - \frac{1}{2}\right)\tau\right\}\right] \\ &\geq \exp\{-(1 + 2\lambda)\}E_Q\left[\exp\left\{\left[\left(2\lambda - \frac{1}{2}\right)\tau\right]\right\}\right] = \infty. \end{aligned}$$

This implies that $a(M) \leq \frac{1}{4} + \frac{\pi^2}{16}$. Thus $d_1(M, L_\infty) \geq \frac{4}{4 + \pi^2}$ by the left-hand side inequality in Theorem 2.8. On the other hand, it is easy to see that $d_1(M, L_\infty) \leq 16$.

Next, we shall make a research into the existence of a continuous decreasing function $\Phi : (1, \infty) \mapsto (0, \infty)$ which satisfies the implication

(3.20) $d_2(M, L_\infty) < \Phi(p) \Longrightarrow \mathcal{E}(M) \in (R_p)$

for each $p > 1$. According to Theorems 3.1 and 3.8, we should proceed on the assumption that $\Phi(1 + 0) = \infty$ and $\Phi(\infty) = 0$. Intuitively speaking, it is our claim that $\mathcal{E}(M)$ must satisfy a stronger reverse Hölder's inequality as M approaches L_∞. Unfortunately, this question is still unsettled, but we have obtained an interesting partial answer which practically guarantees the validity of our claim. To see this, we set

$$\Phi(p) = \left\{1 + \frac{1}{p^2}\log\frac{2p - 1}{2(p - 1)}\right\}^{1/2} - 1 \quad (1 < p < \infty),$$

which is the same function given in (3.4). It is obviously a continuous decreasing function such that $\Phi(1 + 0) = \infty$ and $\Phi(\infty) = 0$. Recall that if $\|M\|_{BMO_2} < \Phi(p)$, then $\mathcal{E}(M)$ satisfies (R_p) by Theorem 3.1. This is a key point in our investigation.

Theorem 3. 9. *Let L_∞^K denote the class of all martingales bounded by a positive constant K and let $1 < p < \infty$. If $d_2(M, L_\infty^K) < e^{-K}\Phi(p)$, then $\mathcal{E}(M)$ has (R_p).*

Proof. By the assumption, $\|M - N\|_{BMO_2} < e^{-K}\Phi(p)$ for some $N \in L_\infty^K$. Let now $d\hat{P} = \mathcal{E}(N)_\infty dP$, which is obviously a probability measure. We set $X = N - M$. According to Theorem 1.8, $\hat{X} = M - N - \langle M - N, N \rangle$ is a martingale with respect to $\hat{P}$ such that $\langle \hat{X} \rangle = \langle X \rangle$, and by the definition of the conditional expectation we have

$$
\begin{aligned}
&\hat{E}[\langle \hat{X} \rangle_\infty - \langle \hat{X} \rangle_T | \mathcal{F}_T] \\
&= E\left[(\langle X \rangle_\infty - \langle X \rangle_T) \exp\left\{ (N_\infty - N_T) - \frac{1}{2}(\langle N \rangle_\infty - \langle N \rangle_T) \right\} \Big| \mathcal{F}_T \right] \\
&\leq e^{2K}\|X\|_{BMO_2}^2 < \Phi(p)^2,
\end{aligned}
$$

that is, $\|\hat{X}\|_{BMO_2(\hat{P})} < \Phi(p)$. Then, according to Theorem 3.1, the exponential martingale $\mathcal{E}(\hat{X})$ satisfies the reverse Hölder inequality

$$
\hat{E}[\mathcal{E}(\hat{X})_\infty^p | \mathcal{F}_T] \leq K_{p,\hat{X}} \mathcal{E}(\hat{X})_T^p.
$$

On the other hand, since $\langle M \rangle = \langle M - N \rangle + 2\langle M - N, N \rangle + \langle N \rangle$, we have

$$
\begin{aligned}
\mathcal{E}(M) &= \exp\left\{ (M - N - \langle M - N, N \rangle) - \frac{1}{2}\langle M - N \rangle \right\} \exp\left(N - \frac{1}{2}\langle N \rangle \right) \\
&= \mathcal{E}(\hat{X})\mathcal{E}(N).
\end{aligned}
$$

Therefore, for every stopping time T we find

$$
\begin{aligned}
E[\mathcal{E}(M)_\infty^p | \mathcal{F}_T] &= E[\{\mathcal{E}(M)_\infty / \mathcal{E}(M)_T\}^p | \mathcal{F}_T]\mathcal{E}(M)_T^p \\
&\leq \exp\{2(p-1)K\}\hat{E}[\{\mathcal{E}(\hat{X})_\infty / \mathcal{E}(\hat{X})_T\}^p | \mathcal{F}_T]\mathcal{E}(M)_T^p \\
&\leq \exp\{2(p-1)K\}K_{p,\hat{X}}\mathcal{E}(M)_T^p,
\end{aligned}
$$

which completes the proof. □

The next example indicates that Theorem 3.9 is rigidly unsatisfactory.

Example 3.5. Let $M_t = B_{t\wedge 1}$ $(0 \leq t < \infty)$, where B is a one dimensional Brownian motion with $B_0 = 0$. Then $M \in \bar{L}_\infty$ clearly. However $d_1(M, L_\infty^K) > 0$ for every $K > 0$, because for every $N \in L_\infty^K$ we have

$$
\begin{aligned}
\|M - N\|_{BMO_1} &\geq E[|B_1 - N_\infty|] \\
&\geq E[|B_1| - |N_\infty| : |B_1| \geq 2K] \\
&\geq KP(|B_1| \geq 2K) > 0.
\end{aligned}
$$

Further, it follows from Example 3.5 that the converse statement in Theorem 3.9 fails. Now we give a variant of Theorem 3.9.

Theorem 3. 10. *Let $1 < p < \infty$. If there exists $N \in \bar{L}_\infty$ such that $\langle M-N, N \rangle = 0$ and $\|M - N\|_{BMO_2} < \Phi(p)$, then $\mathcal{E}(M)$ satisfies (R_p).*

Proof.   Since the function Φ is continuous, $\|M - N\|_{BMO_2} < \Phi(u)$ for some $u > p$. Then $\mathcal{E}(M - N)$ has (R_u) by Theorem 3.1, and $\mathcal{E}(N)$ satisfies (R_r) for all $r > 1$ by theorem 3.8. Furthermore, from the assumption $\langle M - N, N \rangle = 0$ it follows that $\mathcal{E}(M) = \mathcal{E}(M - N)\mathcal{E}(N)$. Hence, applying the Hölder inequality with the exponents $\alpha = u/p$ and $\beta = \alpha/(\alpha - 1)$ gives

$$
\begin{aligned}
E[\mathcal{E}(M)_\infty^p | \mathcal{F}_T] &= E[\{\mathcal{E}(M)_\infty/\mathcal{E}(M)_T\}^p | \mathcal{F}_T] \mathcal{E}(M)_T^p \\
&\leq E[\{\mathcal{E}(M - N)_\infty/\mathcal{E}(M - N)_T\}^u | \mathcal{F}_T]^{1/\alpha} \\
&\quad \cdot E[\{\mathcal{E}(N)_\infty/\mathcal{E}(N)_T\}^{p\beta} | \mathcal{F}_T]^{1/\beta} \mathcal{E}(M)_T^p \\
&\leq C_{p,M,N} \mathcal{E}(M)_T^p.
\end{aligned}
$$

Thus the proof is complete. $\square$

Observe that if M is a continuous martingale, then there is a stopping time T satisfying $M^T \in L_\infty$. Let $\Re(M)$ be the class of these kind of martingales. Then we can obtain the following.

Corollary 3. 5.   *Let* $1 < p < \infty$. *If* $d_2(M, \Re(M)) < \Phi(p)$, *then* $\mathcal{E}(M)$ *has* (R_p).

Proof.   Since $\langle M - M^T, M^T \rangle = 0$ for every stopping time T, the conclusion follows at once from Theorem 3.10. $\square$

3.5   The class H_∞ and the (A_p) condition

Unlike the (R_p) condition, the (A_p) condition is remotely related to the distance between M and H_∞. We first give an example which substantiates this view.

Example 3.6.   Let $M = B^\tau$ where $\tau = \inf\{t : |B_t| = 1\}$. Then $M \in L_\infty$ clearly and so, according to Theorem 3.8, the exponential martingale $\mathcal{E}(\lambda M)$ satisfies all (R_p) for any λ. However, $\mathcal{E}(M)$ does not satisfy (A_p) for p with $1 < p \leq 1 + \frac{4}{\pi^2}$. In fact, since $\frac{1}{2(p-1)} \geq \frac{\pi^2}{8}$ for $1 < p \leq 1 + \frac{4}{\pi^2}$ and $E[\exp(\alpha\tau)] = \infty$ for $\alpha \geq \frac{\pi^2}{8}$, we find

$$
E\left[\{1/\mathcal{E}(M)_\infty\}^{1/(p-1)}\right] \geq \exp(-\frac{1}{p-1}) E\left[\exp(\frac{\tau}{2(p-1)})\right] = \infty,
$$

which implies that for $1 < p \leq 1 + \frac{4}{\pi^2}$ the (A_p) condition fails.

The aim of this section is to claim that H_∞ must be closely connected with the (A_p) condition. Especially, I conjecture that $M \in \bar{H}_\infty$ if and only if both $\mathcal{E}(M)$ and $\mathcal{E}(-M)$ satisfy all (A_p). Unfortunately, we still have not obtained a satisfactory conclusion. So we consider the martingale $q(M) = E[\langle M \rangle_\infty | \mathcal{F}.] - E[\langle M \rangle_\infty | \mathcal{F}_0]$ instead of M, and we shall show that H_∞ has a close relation with the (A_p) condition through the martingale $q(M)$. Note that $\|q(M)\|_{BMO_1} \leq 2\|M\|_{BMO_2}^2$ and that $M \in H_\infty$ if and only if $q(M) \in L_\infty$. The main result here is that both $\mathcal{E}(M)$ and $\mathcal{E}(-M)$ satisfy all (A_p) if and only if $q(M) \in \bar{L}_\infty$. To prove this, we need the following elementary results.

Lemma 3. 1.   *Let $M \in BMO$. Then for every $\lambda > 0$ and every stopping time T we have*

(i)
$$E[\exp\{\lambda(\langle M\rangle_\infty - \langle M\rangle_T)\}|\mathcal{F}_T]$$
$$\leq \exp(\lambda\|M\|^2_{BMO_2})E[\exp\{\lambda|\langle M\rangle_\infty - E[\langle M\rangle_\infty|\mathcal{F}_T]|\}|\mathcal{F}_T]$$

(ii)
$$E[\exp\{\lambda|\langle M\rangle_\infty - E[\langle M\rangle_\infty|\mathcal{F}_T]|\}|\mathcal{F}_T]$$
$$\leq \exp(\lambda\|M\|^2_{BMO_2})E[\exp\{\lambda(\langle M\rangle_\infty - \langle M\rangle_T)\}|\mathcal{F}_T].$$

Proof.   By the definition of the norm $\|M\|_{BMO_2}$ we have for every $\lambda > 0$

$$E[\exp\{\lambda(\langle M\rangle_\infty - \langle M\rangle_T)\}|\mathcal{F}_T]$$
$$\leq E[\exp\{\lambda|\langle M\rangle_\infty - E[\langle M\rangle_\infty|\mathcal{F}_T]|\} \exp\{\lambda E[\langle M\rangle_\infty - \langle M\rangle_T|\mathcal{F}_T]\}|\mathcal{F}_T]$$
$$\leq \exp(\lambda\|M\|^2_{BMO_2})E[\exp(\lambda|\langle M\rangle_\infty - E[\langle M\rangle_\infty|\mathcal{F}_T]|)|\mathcal{F}_T].$$

Thus (i) holds. The proof of (ii) is similar, and so we omit it. □

Theorem 3. 11.   *Let $M \in BMO$ and $1 < p < \infty$.*

(i) *If $d_1(q(M), L_\infty) < \frac{1}{2}(\sqrt{p} - 1)^2$, then both $\mathcal{E}(M)$ and $\mathcal{E}(-M)$ satisfy (A_p).*
(ii) *Conversely, if both $\mathcal{E}(M)$ and $\mathcal{E}(-M)$ satisfy (A_p), then*

$$d_1(q(M), L_\infty) < 8(p - 1).$$

Proof.   We first show (i). If $d_1(q(M), L_\infty) < (\sqrt{p} - 1)^2/2$, then, according to the left-hand side inequality of Theorem 2.8, we have

$$a(q(M)) > \frac{1}{2(\sqrt{p} - 1)^2}$$

and so by (i) in Lemma 3.1

$$E\left[\exp\left\{\frac{1}{2(\sqrt{p} - 1)^2}(\langle M\rangle_\infty - \langle M\rangle_T)\right\}\Big|\mathcal{F}_T\right] \leq C_p.$$

Let now $r = \sqrt{p} + 1$. The exponent conjugate to r is $s = (\sqrt{p} + 1)/\sqrt{p}$. Since $\{s(\sqrt{p} - 1)^2\}^{-1} - r(p - 1)^{-2} = (p - 1)^{-1}$, we have

$$\{\mathcal{E}(M)_T/\mathcal{E}(M)_\infty\}^{\frac{1}{p-1}} = \exp\left\{-\frac{1}{p-1}(M_\infty - M_T) - \frac{r}{2(p-1)^2}(\langle M\rangle_\infty - \langle M\rangle_T)\right\}$$
$$\cdot \exp\left\{\frac{1}{2s(\sqrt{p} - 1)^2}(\langle M\rangle_\infty - \langle M\rangle_T)\right\}.$$

Then, applying Hölder's inequality with the exponents r and s gives

$$E\left[\{\mathcal{E}(M)_T/\mathcal{E}(M)_\infty\}^{\frac{1}{p-1}}\Big|\mathcal{F}_T\right] \leq E\left[\mathcal{E}\left(-\frac{r}{p-1}M\right)_\infty\Big/\mathcal{E}\left(-\frac{r}{p-1}M\right)_T\Big|\mathcal{F}_T\right]^{1/r}$$
$$\cdot E\left[\exp\left\{\frac{1}{2(\sqrt{p} - 1)^2}(\langle M\rangle_\infty - \langle M\rangle_T)\right\}\Big|\mathcal{F}_T\right]^{1/s}.$$

The first conditional expectation on the right hand side equals 1, because $\mathcal{E}\left(-\frac{r}{p-1}M\right)$ is a uniformly integrable martingale. The second term is dominated by some constant C_p as is stated above. In the same way we can verify that $\mathcal{E}(-M)$ satisfies the (A_p) condition.

To show (ii), it suffices to apply

$$E\left[\exp\left\{\frac{1}{2(p-1)}(\langle M\rangle_\infty - \langle M\rangle_T)\right\}\Big|\mathcal{F}_T\right]$$

$$\leq E\left[\{\mathcal{E}(M)_T/\mathcal{E}(M)_\infty\}^{\frac{1}{p-1}}\Big|\mathcal{F}_T\right]^{1/2}E[\{\mathcal{E}(-M)_T/\mathcal{E}(-M)_\infty\}^{\frac{1}{p-1}}|\mathcal{F}_T]^{1/2},$$

which follows from the Schwarz inequality. This inequality implies that

$$E\left[\exp\left\{\frac{1}{2(p-1)}(\langle M\rangle_\infty - \langle M\rangle_T)\right\}\Big|\mathcal{F}_T\right] \leq C_p$$

whenever both $\mathcal{E}(M)$ and $\mathcal{E}(-M)$ have (A_p). Then $a(q(M)) \geq \dfrac{1}{2(p-1)}$ by (ii) in Lemma 3.1. On the other hand, if $\mathcal{E}(M)$ has (A_p), then it satisfies $(A_{p-\epsilon})$ for some ϵ with $0 < \epsilon < p-1$ by Corollary 3.3. Thus $a(q(M)) > \dfrac{1}{2(p-1)}$ in fact, so that $d_1(q(M), L_\infty) < 8(p-1)$ by the right-hand side inequality of Theorem 2.8. This completes the proof.   □

As a corollary, we obtain the following.

Theorem 3. 12.   *In order that both $\mathcal{E}(M)$ and $\mathcal{E}(-M)$ satisfy all (A_p), a necessary and sufficient condition is that $q(M) \in \bar{L}_\infty$.*

It is probably that $q(M) \in \bar{L}_\infty$ if and only if $M \in \bar{II}_\infty$. To verify the "if" part is not difficult, but the converse remains to be proved. Here we will only remark that if $q(M) \in \bar{L}_\infty$, then $M \in \bar{L}_\infty$. An application of the Schwarz inequality yields that for every $\lambda > 0$

$$\begin{aligned}
E[\exp\{\lambda(M_\infty - M_T)\}|\mathcal{F}_T] \\
= E[\{\mathcal{E}(2\lambda M)_\infty/\mathcal{E}(2\lambda M)_T\}^{1/2}\exp\{\lambda^2(\langle M\rangle_\infty - \langle M\rangle_T)\}|\mathcal{F}_T] \\
\leq E[\mathcal{E}(2\lambda M)_\infty/\mathcal{E}(2\lambda M)_T|\mathcal{F}_T]^{1/2}E[\exp\{2\lambda^2(\langle M\rangle_\infty - \langle M\rangle_T)\}|\mathcal{F}_T]^{1/2} \\
\leq E[\exp\{2\lambda^2(\langle M\rangle_\infty - \langle M\rangle_T)\}|\mathcal{F}_T]^{1/2}.
\end{aligned}$$

The same argument works if M is replaced by $-M$. Then for all $\lambda > 0$

$$E[\exp\{\lambda|M_\infty - M_T|\}|\mathcal{F}_T] \leq 2E[\exp\{2\lambda^2(\langle M\rangle_\infty - \langle M\rangle_T)\}|\mathcal{F}_T]^{1/2}.$$

Therefore, from Theorem 2.8 and Lemma 3.1 it follows immediately that if $q(M) \in \bar{L}_\infty$, then $M \in \bar{L}_\infty$.

We end this section with a very interesting remark on the distance between M and H_∞.

Theorem 3. 13.   *Let $1 < p < \infty$ and $p^{-1} + q^{-1} = 1$. If $d_2(M, H_\infty) < \Phi(p)$ where Φ is the function given in (3.4), then $\mathcal{E}(M)$ satisfies both (R_p) and (A_q).*

Proof. By the assumption, $\|M - N\|_{BMO_2} < \Phi(p)$ for some $N \in H_\infty$. Consider now the new probability measure $dQ = \mathcal{E}(M - N)_\infty dP$ and set $\tilde{N} = N - \langle M - N, N \rangle$. Then $\tilde{N} \in BMO(Q)$ and $\langle \tilde{N} \rangle = \langle N \rangle$ by Theorem 1.8. Further, we have

$$(3.21) \qquad \mathcal{E}(M) = \mathcal{E}(M - N)\mathcal{E}(\tilde{N}),$$

where $\mathcal{E}(\tilde{N})$ is the exponential martingale under under dQ.

We first verify that $\mathcal{E}(M)$ satisfies (R_p). By the definition of H_∞, $\langle N \rangle$ is bounded, so that $\tilde{N} \in H_\infty(Q) \subset \bar{L}_\infty(Q)$. Therefore, it follows from Theorem 3.8 that $\mathcal{E}(\tilde{N})$ satisfies all (R_r) under dQ. Then, noticing (3.21) and applying the Hölder inequality with exponents r and $s = \frac{r}{r-1}$, we find

$$
\begin{aligned}
E[\{\mathcal{E}(M)_\infty/\mathcal{E}(M)_T\}^p | \mathcal{F}_T] &= E[\{\mathcal{E}(M - N)_\infty/\mathcal{E}(M - N)_T\}^{p - 1/r} \\
&\quad \cdot \{\mathcal{E}(M - N)_\infty/\mathcal{E}(M - N)_T\}^{1/r} \{\mathcal{E}(\tilde{N})_\infty/\mathcal{E}(\tilde{N})_T\}^p | \mathcal{F}_T] \\
&\le E[\{\mathcal{E}(M - N)_\infty/\mathcal{E}(M - N)_T\}^{(p - 1/r)s} | \mathcal{F}_T]^{1/s} \\
&\quad \cdot E_Q[\{\mathcal{E}(\tilde{N})_\infty/\mathcal{E}(\tilde{N})_T\}^{pr} | \mathcal{F}_T]^{1/r} \\
&\le C_{p,r} E[\{\mathcal{E}(M - N)_\infty/\mathcal{E}(M - N)_T\}^{(p - 1/r)s} | \mathcal{F}_T]^{1/s}.
\end{aligned}
$$

Since $p < (p - 1/r)s = (pr - 1)/(r - 1) \to p$ as $r \to \infty$, there is some $r > 1$ such that

$$\|M - N\|_{BMO_2} < \Phi((p - 1/r)s).$$

Then, according to Theorem 3.1, $\mathcal{E}(M - N)$ satisfies the $(R_{(p-1/r)s})$ condition. Thus $\mathcal{E}(M)$ has (R_p).

We are now going to prove that $\mathcal{E}(M)$ satisfies (A_q). An elementary calculation shows that $\Phi(p) < \sqrt{q} - 1$, so that

$$\|M - N\|_{BMO_2} < \sqrt{q} - 1.$$

Since $\langle M \rangle_t - \langle M \rangle_s \le 2\{(\langle M - N \rangle_t - \langle M - N \rangle_s) + (\langle N \rangle_t - \langle N \rangle_s)\}$ for $s < t$, an application of Theorem 2.2 yields

$$
\begin{aligned}
E\left[\exp\left\{\frac{1}{2(\sqrt{q} - 1)^2}(\langle M \rangle_\infty - \langle M \rangle_T)\right\} \middle| \mathcal{F}_T\right] \\
\le C_q E\left[\exp\left\{\frac{1}{(\sqrt{q} - 1)^2}(\langle M - N \rangle_\infty - \langle M - N \rangle_T)\right\} \middle| \mathcal{F}_T\right] \\
\le C_q \left\{1 - \frac{1}{(\sqrt{q} - 1)^2}\|M - N\|_{BMO_2}^2\right\}^{-1}.
\end{aligned}
$$

Let now $r = \sqrt{q} + 1$ and $s = \frac{\sqrt{q}+1}{\sqrt{q}}$. Then by the Hölder inequality we have

$$
\begin{aligned}
E[\{\mathcal{E}(M)_T/\mathcal{E}(M)_\infty\}^{1/(q-1)} | \mathcal{F}_T] \\
= E\left[\exp\left\{-\frac{1}{q-1}(M_\infty - M_T) - \frac{r}{2(q-1)^2}(\langle M \rangle_\infty - \langle M \rangle_T)\right\} \right. \\
\left. \cdot \exp\left\{\frac{1}{2s(\sqrt{q} - 1)^2}(\langle M \rangle_\infty - \langle M \rangle_T)\right\} \middle| \mathcal{F}_T\right] \\
\le E\left[\mathcal{E}\left(-\frac{r}{q-1}M\right)_\infty \middle/ \mathcal{E}\left(-\frac{r}{q-1}M\right)_T \middle| \mathcal{F}_T\right]^{1/r} \\
\cdot E\left[\exp\left\{\frac{1}{2(\sqrt{q} - 1)^2}(\langle M \rangle_\infty - \langle M \rangle_T)\right\} \middle| \mathcal{F}_T\right]^{1/s}.
\end{aligned}
$$

The first conditional expectation in the last expression equals to 1, and the second one is bounded by some constant as is remarked above. Thus the proof is complete.

$\square$

Combining Theorems 3.12 and 3.13 shows that $\bar{H}_\infty \subset \{M : q(M) \in \bar{L}_\infty\}$. As mentioned before, I conjecture that these classes must be equal. A key to verify it is to establish the inequality

$$b(M) \leq \frac{C}{d_2(M, H_\infty)},$$

where $b(M)$ denotes the supremum of the set of b for which

$$\sup_T \|E[\exp\{b^2(\langle M\rangle_\infty - \langle M\rangle_T)\}|\mathcal{F}_T]\|_\infty < \infty.$$

But this remains to be proved.

3.6   Weighted norm inequalities

Let $1 < p < \infty$ and $0 \leq w \in L^1_{loc}(R^n)$. Then, as is stated in Section 2.3, the inequality

$$\int_{R^n} f^*(x)^p w(x)dx \leq C_p \int_{R^n} |f(x)|^p w(x)dx$$

is valid for all $f \in L^p(w(x)dx)$ if and only if w satisfies the Muckenhoupt (A_p) condition

$$\sup_Q \left(\frac{1}{|Q|} \int_Q w(x)dx\right) \left(\frac{1}{|Q|} \int_Q w(x)^{\frac{1}{p-1}} dx\right)^{p-1} < \infty.$$

Here $f^*(x)$ denotes the Hardy-Littlewood maximal function of f. In 1975 C. Watari raised the question of searching the possibility of its extension to the martingales. The aim of this section is to establish several weighted norm inequalities for martingales. For that, let M be a continuous local martingale with $M_0 = 0$, and we assume that $\mathcal{E}(M)$ is a uniformly integrable martingale. We begin with the weighted norm inequality problem for the maximal function. To be concrete, it is the characterization of the weight martingale $\mathcal{E}(M)$ for which the inequality

(3.22) $\hat{E}[(X^*)^p] \leq C_p \hat{E}[|X_\infty|^p]$

holds for all uniformly integrable martingales X, where $1 < p < \infty$ and C_p is a constant independent of X. It goes without saying that the inequality (3.22) for the case where $\mathcal{E}(M) = 1$ is nothing but the Doob inequality.

The following was given by T. Tsuchikura ([81]) in a general setting.

Theorem 3. 14.   *Let $1 \leq p < \infty$. If $\mathcal{E}(M)$ satisfies (A_p), then the inequality*

(W_p) $\lambda^p \hat{P}(X^* > \lambda) \leq C_p \hat{E}[|X_\infty|^p : X^* > \lambda]$ $(0 < \lambda < \infty)$

holds for all uniformly integrable martingales X relative to P.

Proof. Let X be any uniformly integrable martingale relative to P. We first show that the inequality

(3.23) $$|X_T|^p \le C_p \hat{E}[|X_\infty|^p|\mathcal{F}_T] \quad (1 \le p < \infty)$$

holds for any stopping time T, with a positive constant C_p depending only on M and p. When $p = 1$, it follows at once from the definition of (A_1). So, let $1 < p < \infty$. Then

$$\begin{aligned}|X_T|^p &\le E[\{\mathcal{E}(M)_T/\mathcal{E}(M)_\infty\}^{1/p}\{\mathcal{E}(M)_\infty/\mathcal{E}(M)_T\}^{1/p}|X_\infty||\mathcal{F}_T]^p \\ &\le E[\{\mathcal{E}(M)_T/\mathcal{E}(M)_\infty\}^{1/(p-1)}|\mathcal{F}_T]^{p-1}\hat{E}[|X_\infty|^p|\mathcal{F}_T],\end{aligned}$$

by the Hölder inequality, and the first conditional expectation in the last expression is dominated by some constant C_p. This yields (3.23).

Now, let $T_\lambda = \inf\{t \ge 0 : |X_t| > \lambda\}$ for $\lambda > 0$. Then

$$\{X^* > \lambda\} = \{T < \infty\} \subset \{|X_T| \ge \lambda\}$$

and $\{X^* > \lambda\} \in \mathcal{F}_T$ obviously. Thus it follows immediately from (3.23) that

$$\begin{aligned}\lambda^p \hat{P}(X^* > \lambda) &\le \hat{E}[|X_T|^p : X^* > \lambda] \\ &\le C_p \hat{E}[|X_\infty| : X^* > \lambda].\end{aligned}$$

$\square$

Two years later, A. Uchiyama showed in [82] that the converse of Theorem 3.14 is true. We shall explain his result in the following.

Theorem 3. 15. *Let $1 \le p < \infty$. If the inequality (W_p) holds for every uniformly integrable martingale X and every $\lambda > 0$, then $\mathcal{E}(M)$ has (A_p).*

Proof. We first show (3.23), following the idea of C. Doléans-Dade and P. A. Meyer. For that, let X be any uniformly integrable martingale. For each stopping time T and $A \in \mathcal{F}_T$ we set

$$Y_t = E[X_\infty I_A|\mathcal{F}_t] \quad (0 \le t < \infty).$$

It is clear that $Y_t = X_t I_A$ on $\{T \le t\}$ and $Y_\infty = X_\infty I_A$, so that

$$A \cap \{|X_T| > \lambda\} \subset \{Y^* > \lambda\}.$$

Then by (W_p) we have

$$\begin{aligned}\lambda^p \hat{P}(|X_T| > \lambda, A) &\le \lambda^p \hat{P}(Y^* > \lambda) \\ &\le C_p \hat{E}[|Y_\infty|^p] \\ &\le C_p \hat{E}[|X_\infty|^p : A]\end{aligned}$$

for every $\lambda > 0$, with a constant C_p depending only on M and p. This implies that for any $\lambda > 0$

$$\lambda^p I_{\{|X_T|>\lambda\}} \le C_p \hat{E}[|X_\infty|^p|\mathcal{F}_T],$$

which yields (3.23), that is,

$$|X_T|^p \leq C_p \hat{E}[|X_\infty|^p | \mathcal{F}_T].$$

Suppose now that $\mathcal{E}(M)$ satisfies (W_1). If we set $X_t = P(\Lambda|\mathcal{F}_t)$ $(0 \leq t < \infty)$ where $\Lambda \in \mathcal{F}$, then it follows from (3.23) that for any stopping time T

$$\begin{aligned} E[\mathcal{E}(M)_T : \Lambda] &= E[\mathcal{E}(M)_T X_T] \\ &\leq CE[\mathcal{E}(M)_\infty : \Lambda]. \end{aligned}$$

Thus $\mathcal{E}(M)_T \leq C\mathcal{E}(M)_\infty$, which means that $\mathcal{E}(M)$ satisfies (A_1).
Next, we shall deal with the case where $\mathcal{E}(M)$ satisfies (W_p) for $p > 1$. For each stopping time T let

$$X_t = E\left[I_{D \cap \{\mathcal{E}(M)_\infty \geq a\}} \mathcal{E}(M)_\infty^{-\frac{1}{p-1}} \Big| \mathcal{F}_t \right] \quad (0 \leq t < \infty)$$

where $a > 0$ and $D \in \mathcal{F}_T$.
Note that

$$0 \leq X_t \leq a^{-\frac{1}{p-1}}, X_\infty = I_{D \cap \{\mathcal{E}(M)_\infty \geq a\}} \mathcal{E}(M)_\infty^{-\frac{1}{p-1}}$$

and

$$X_T = E\left[I_{\{\mathcal{E}(M)_\infty \geq a\}} \mathcal{E}(M)_\infty^{-\frac{1}{p-1}} \Big| \mathcal{F}_T \right] I_D.$$

Then by (3.23) we have

$$E\left[\mathcal{E}(M)_T E\left[\mathcal{E}(M)_\infty^{-\frac{1}{p-1}} I_{\{\mathcal{E}(M)_\infty \geq a\}} \Big| \mathcal{F}_T \right]^p : D \right]$$
$$\leq C_p E\left[\mathcal{E}(M)_\infty^{-\frac{1}{p-1}} I_{\{\mathcal{E}(M)_\infty \geq a\}} : D \right],$$

which yields

$$\mathcal{E}(M)_T E\left[\mathcal{E}(M)_\infty^{-\frac{1}{p-1}} I_{\{\mathcal{E}(M)_\infty \geq a\}} \Big| \mathcal{F}_T \right]^{p-1} \leq C_p.$$

Letting $a \downarrow 0$, it follows that $\mathcal{E}(M)$ has (A_p). This completes the proof. □

Our first object of this section is to prove the next result which was obtained by M. Izumisawa and N. Kazamaki ([28]).

Theorem 3. 16. *Let $1 < p < \infty$. In order that the inequality*

$$\hat{E}[(X^*)^p] \leq C_p \hat{E}[|X_\infty|^p]$$

holds for all uniformly integrable martingales X relative to P, a necessary and sufficient condition is that $\mathcal{E}(M)$ satisfies (A_p).

Proof. We first show that the condition is necessary. For a uniformly integrable martingale X let

$$Y_t = E[X_\infty I_{\{T < \infty\}} | \mathcal{F}_t] \quad (0 \leq t < \infty)$$

where $T = \inf\{t : |X_t| > \lambda\}$ $(\lambda > 0)$. Then

$$\{X^* > \lambda\} \subset \{Y^* > \lambda\},$$

because $\{X^* > \lambda\} = \{T < \infty\}$ and $Y = X$ on the stochastic interval $[T, \infty[$. Furthermore, by the assumption we find

$$
\begin{aligned}
\lambda^p \hat{P}(X^* > \lambda) &\leq \lambda^p \hat{P}(Y^* > \lambda) \\
&\leq \hat{E}[(Y^*)^p] \\
&\leq C_p \hat{E}[|Y_\infty|^p] \\
&\leq C_p \hat{E}[|X_\infty|^p : T < \infty] \\
&\leq C_p \hat{E}[|X_\infty|^p : X^* > \lambda],
\end{aligned}
$$

which implies (W_p). Thus $\mathcal{E}(M)$ satisfies (A_p) by Theorem 3.15.

Conversely, if $\mathcal{E}(M)$ satisfies (A_p) for $p > 1$, then it satisfies $(A_{p-\varepsilon})$ for some ε with $0 < \varepsilon < p - 1$ by Corollary 3.3. Let now X be any uniformly integrable martingale. Without loss of generality we may assume that $X_\infty \in L^{p-\varepsilon}(\hat{P})$. Then, applying Hölder's inequality with the exponents $p - \varepsilon$ and $\frac{p-\varepsilon}{p-\varepsilon-1}$ gives

$$
\begin{aligned}
|X_T| &\leq E\left[\{\mathcal{E}(M)_\infty/\mathcal{E}(M)_T\}^{\frac{1}{p-\varepsilon}}\{\mathcal{E}(M)_T/\mathcal{E}(M)_\infty\}^{\frac{1}{p-\varepsilon}}|X_\infty| \,\Big|\, \mathcal{F}_T\right] \\
&\leq \hat{E}[|X_\infty|^{p-\varepsilon}|\mathcal{F}_T]^{\frac{1}{p-\varepsilon}} E\left[\{\mathcal{E}(M)_T/\mathcal{E}(M)_\infty\}^{\frac{1}{p-\varepsilon-1}} \,\Big|\, \mathcal{F}_T\right]^{\frac{p-\varepsilon-1}{p-\varepsilon}}
\end{aligned}
$$

for every stopping time T. Thus

$$
X^* \leq C_{p,\varepsilon} \sup_t \hat{E}[|X_\infty|^{p-\varepsilon}|\mathcal{F}_t]^{1/(p-\varepsilon)}
$$

where the constant $C_{p,\varepsilon}$ depends only on p and ε. Then by Doob's inequality we have

$$
\begin{aligned}
\hat{E}[(X^*)^p] &\leq C_{p,\varepsilon} \hat{E}\left[\sup_t \hat{E}[|X_\infty|^{p-\varepsilon}|\mathcal{F}_t]^{\frac{p}{p-\varepsilon}}\right] \\
&\leq C_{p,\varepsilon} \hat{E}[|X_\infty|^p],
\end{aligned}
$$

which completes the proof. $\qquad\square$

Observe that $\mathcal{E}(\hat{M})$ satisfies the (A_p) condition with respect to $\hat{P}$ if and only if $\mathcal{E}(M)$ satisfies the reverse Hölder inequality (R_q) where $p^{-1} + q^{-1} = 1$. Thus, looking at Theorem 3.16 from another angle, we get the following.

Corollary 3. 6. *Let $M \in BMO$ and $1 < p < \infty$. The inequality*

$$
E\left[\sup_t |X_t - \langle X, M\rangle_t|^p\right] \leq C_p \sup_t E[|X_t - \langle X, M\rangle_t|^p]
$$

is valid for all martingales X if and only if $\mathcal{E}(M)$ satisfies the reverse Hölder inquality (R_q).

Next, we shall mention the weighted norm inequalities of Burkholder -Davis-Gundy's type which were established in a general setting by T. Sekiguchi ([74]) and independently by A. Bonami and D. Lépingle ([3]).

Lemma 3. 2. *Let L be a continuous local martingale, H a predictable process and A a continuous increasing process. If $|AL| \leq 1$ and $|H| \leq A$, then $\|H \circ L\|_{BMO_2} \leq \sqrt{2}$.*

Proof.   The usual stopping time argument enables us to assume that $L, L^2 - \langle L \rangle$ and $H \circ L$ are uniformly integrable martingales. Let $N = H \circ L$ for simplicity. Then the integration by parts formula gives

$$
\begin{aligned}
\langle N \rangle_\infty - \langle N \rangle_T &= \int_T^\infty H_s^2 d\langle L \rangle_s \\
&\le \int_T^\infty A_s^2 d\langle L \rangle_s \\
&\le \int_T^\infty (\langle L \rangle_\infty - \langle L \rangle_s) dA_s^2 + (\langle L \rangle_\infty - \langle L \rangle_T) A_T^2,
\end{aligned}
$$

and so, since $E[\langle L \rangle_\infty - \langle L \rangle_s | \mathcal{F}_s] \le E[L_\infty^2 | \mathcal{F}_s]$, we have

$$
E[\langle N \rangle_\infty - \langle N \rangle_T | \mathcal{F}_T] \le 2E[L_\infty^2 A_\infty^2 | \mathcal{F}_T] \le 2.
$$

Thus the lemma is proved.   □

Theorem 3. 17.   *If $M \in BMO(P)$, then the inequalities*

$$
(3.24) \qquad\qquad c\hat{E}[X^*] \le \hat{E}[\langle X \rangle_\infty^{1/2}] \le C\hat{E}[X^*]
$$

hold for all martingales X relative to P, where the choice of c and C depends only on M.

Proof.   We first establish the left-hand inequality. If X is a martingale, then $\hat{X} = \langle X, M \rangle - X$ is a local martingale relative to $\hat{P}$ by Theorem 1.8. Since $\langle X, M \rangle = \langle \hat{X}, \hat{M} \rangle$, we have

$$
X^* \le \int_0^\infty |d\langle \hat{X}, \hat{M} \rangle_t| + \hat{X}^*.
$$

Recall that $\hat{M} \in BMO(\hat{P})$ if $M \in BMO$ (cf. Theorem 3.3). Then, combining Fefferman's inequality and Davis inequality shows

$$
\hat{E}[X^*] \le \sqrt{2} \|\hat{M}\|_{BMO_2(\hat{P})} \|\hat{X}\|_{H_1(\hat{P})} + \hat{E}[\hat{X}^*] \le C_M \hat{E}[\langle X \rangle_\infty^{1/2}],
$$

where $C_M = \sqrt{2} \|\hat{M}\|_{BMO_2(\hat{P})} + 4\sqrt{2}$.

Next, in order to verify the right-hand inequality, we may assume that $\hat{E}[X^*] \le 1$. Then

$$
\begin{aligned}
\hat{E}[\langle X \rangle_\infty^{1/2}] &\le \hat{E}[X^*]^{1/2} \hat{E}\left[\frac{\langle X \rangle_\infty}{X^*}\right]^{1/2} \\
&\le \left\{ 1 + \left| \hat{E}\left[(X_\infty^2 - \langle X \rangle_\infty) \frac{1}{X^*} \right] \right| \right\}^{1/2}.
\end{aligned}
$$

Here we may assume further that $1/X^*$ is bounded. For $t \ge 0$ let

$$
Y_t = \hat{E}[1/X^* | \mathcal{F}_t].
$$

Then

$$
\|X \circ Y\|_{BMO_2(\hat{P})} \le \sqrt{2}
$$

by Lemma 3.2. Since $X^2 - \langle X \rangle = 2X \circ X$ and $X = \langle \hat{X}, \hat{M} \rangle - \hat{X}$, applying Fefferman 's inequality gives

$$|\hat{E}[(X_\infty^2 - \langle X \rangle_\infty)Y_\infty]|$$

$$= 2 \left| \hat{E} \left[\left(\int_0^\infty X_s dX_s \right) Y_\infty \right] \right|$$

$$\leq 2|\hat{E}[(X \circ \hat{X})_\infty Y_\infty]| + 2\hat{E} \left[Y_\infty \int_0^\infty |X_s||d\langle \hat{X}, \hat{M} \rangle_s| \right]$$

$$\leq 2|\hat{E}[\hat{X}_\infty (X \circ Y)_\infty]| + 2\hat{E} \left[\int_0^\infty \frac{|X_s|}{X^*} |d\langle \hat{X}, \hat{M} \rangle_s| \right]$$

$$\leq 2\sqrt{2}\|\hat{X}\|_{H_1(\hat{P})} \|X \circ Y\|_{BMO_2(\hat{P})} + 2\sqrt{2}\|\hat{X}\|_{H_1(\hat{P})} \|\hat{M}\|_{BMO_2(\hat{P})}$$

$$\leq \left(4 + 2\sqrt{2}\|\hat{M}\|_{BMO_2(\hat{P})} \right) \|\hat{X}\|_{H_1(\hat{P})}.$$

Thus we have

$$\hat{E}[\langle X \rangle_\infty^{1/2}] \leq \{1 + C_M \hat{E}[\langle X \rangle_\infty^{1/2}]\}^{1/2},$$

form which it follows that

$$\hat{E}[\langle X \rangle_\infty^{1/2}] \leq \frac{1}{2} \left(C_M + \sqrt{C_M^2 + 4} \right).$$

This completes the proof. □

In [75] T. Sekiguchi proved that the converse of Theorem 3.17 holds as follows.

Theorem 3. 18. *Assume that $\mathcal{E}(M)$ is a uniformly integrable martingale. If the inequality*

(3.25) $$\hat{E}[X^*] \leq C\hat{E}[\langle X \rangle_\infty^{1/2}]$$

is valid for all martingales X, then M belongs to the class $BMO(P)$.

Proof. Let $X' \in H_1(\hat{P})$ and let $X = \langle X', \hat{M} \rangle - X'$. Then, according to Theorem 1.8, X is a local martingale with respect to P such that $\langle X \rangle = \langle X' \rangle$ and $X' = \phi(X)$ where ϕ is the Girsanov transformation. Let now $D = |d\langle X, M \rangle|/d\langle X, M \rangle$. Since D is a predictable process such that $D^2 = 1$ and $|\langle X, M \rangle_t| = \int_0^t D_s d\langle X, M \rangle_s$, combining (3.24) and the Davis inequality gives

$$\hat{E} \left[|\langle X', \hat{M} \rangle_\infty| \right] = \hat{E}[|\langle X, M \rangle_\infty|]$$

$$= \hat{E}[\langle D \circ X, M \rangle_\infty]$$

$$\leq \hat{E} \left[(\widehat{D \circ X})^* \right] + \hat{E}[X^*]$$

$$\leq 4\sqrt{2}\hat{E} \left[\langle D \circ X \rangle_\infty^{1/2} \right] + C\hat{E} \left[\langle X \rangle_\infty^{1/2} \right]$$

$$\leq C\hat{E} \left[\langle X' \rangle_\infty^{1/2} \right]$$

for every $X' \in H_1(\hat{P})$. Then $\hat{M} \in BMO(\hat{P})$ by (2.16) in Theorem 2.7, and so it follows at once from Theorem 3.3 that $M \in BMO(P)$. Thus the proof is complete. □

Finally, we shall give a generalization of the two-sided inequality (3.24).

Definition 3. 2.   *Let $F : \mathbb{R}_+ \to \mathbb{R}_+$ be a non-decreasing function with $F(0) = 0$. Then F is called a moderately increasing function if it satisfies the growth condition*

$$F(2x) \le cF(x) \text{for all } x \in R_+.$$

For instance, x^p $(0 < p < \infty)$ and $(x + 1)\log(x + 1)$ are moderately increasing functions, while e^x isn't.

The next inequality follows from (3.24) by the similar method as the proof of the Burkholder-Davis-Gundy inequality, and so we omit its proof.

Theorem 3. 19.   *Assume that $M \in BMO(P)$. If F is a moderately increasing function, then the inequality*

$$c\hat{E}[F(X^*)] \le \hat{E}[F(\langle X \rangle_\infty^{1/2})] \le C\hat{E}[F(X^*)]$$

is valid for all martingales X, where the choice of c and C depends only on M and the growth constant of F.

3.7   Some ratio inequalities

In this section we shall establish various ratio inequalities for martingales. Let X be a continuous local martingale and let $(L_t^a)_{t \ge 0, a \in R}$ be the family of its local times : that is, L^a is the unique continuous increasing process A such that $|X - a| - A$ is a local martingale. For $0 \le t < \infty$ let $L_t^* = \sup_{a \in \mathbb{R}} L_t^a$. Then it was shown in [1] by M. T. Barlow and M. Yor that the process L^* is also continuous and increasing.

Let now Y_1 and Y_2 be any two of the three random variables X^*, $\langle X \rangle_\infty^{1/2}$, L_∞^*, and let $\Phi : [0, \infty] \to [0, \infty]$ be an increasing function. The aim of this section is to study the problem : does there exist a constant $C > 0$, depending only on p and Φ, such that the ratio inequality

$$(3.26) \qquad E[Y_1^p \Phi(Y_1/Y_2)] \le CE[Y_1^p] \quad (0 < p < \infty)$$

is valid for all martingales X ? It is possible to establish the ratio inequality for non-continuous martingales, but for convenience' sake we continue to assume the sample continuity of all martingales.

The above inequality for the case where $\Phi(x) = x^r$ $(r > 0)$ was obtained in 1982 by R. F. Gundy ([24]) and independently by M. Yor ([88]). Quite recently, we have improved their results to the case where $\Phi(x) = \exp(\alpha x)$ for $\alpha > 0$ (see [42]). Note that the inequality (3.26) does not necessarily hold for any increasing function Φ. Before proving our result, we shall exemplify this.

Example 3.7. Let $B = (B_t, \mathcal{F}_t)$ be a one-dimensional Brownian motion starting at 0, and we set $X_t = B_{t \wedge 1}$, $\Phi(x) = \exp(x^2/2)$. Then, noticing $\langle X \rangle_\infty = 1$ we have

$$\sqrt{e} + E[X^{*^p}\Phi(X^*/\langle X \rangle_\infty^{1/2}) : X^* > 1] \ge E\left[\exp(\frac{1}{2}B_1^2)\right] = \infty.$$

Since $X^* \in L_p$ for every $p > 0$, (3.26) fails if $Y_1 = X^*$ and $Y_2 = \langle X \rangle_\infty^{1/2}$.

Next we give an example such that (3.26) fails in the case where if $Y_1 = \langle X \rangle_\infty^{1/2}$ and

$Y_2 = X^*$.

Example 3.8. Let X be the martingale defined by $X_t = B_{t \wedge \tau}$ where $\tau = \inf\{t : |B_t| = 1\}$. It is clear that $X^* = 1$ and $\langle X \rangle_\infty = \tau$. From the Burkholder-Davis-Gundy inequality it follows immediately that $\langle X \rangle_\infty^{1/2} \in L_p$ for every $p > 0$. Let now $\Phi(x) = \exp(\frac{1}{8}\pi^2 x^2)$. Then

$$E\left[\exp\left(\frac{1}{8}\pi^2\tau\right)\right] \leq \exp\left(\frac{1}{8}\pi^2\right) + E[\langle X \rangle_\infty^{p/2}\Phi(\langle X \rangle_\infty^{1/2}/X^*); \langle X \rangle_\infty > 1]$$

and by Lemma 1.3 the expectation on the left-hand side is infinite. Thus

$$E[\langle X \rangle_\infty^{p/2}\Phi(\langle X \rangle_\infty^{1/2}/X^*)] = \infty$$

for every $p > 1$.

One of the ratio inequalities we will prove is that for every $0 < p < \infty$ and every $0 \leq \alpha < \infty$ the inequality

$$E[X^{*p}\exp(\alpha X^*/\langle X \rangle_\infty^{1/2})] \leq C_{\alpha,p}E[X^{*p}]$$

is valid for all continuous martingales X. But the essential result of this section is instead the following.

Theorem 3. 20. *Let U and V be two right-continuous increasing processes adapted to the filtration $(\mathcal{F}_t)$. If there exists a constant $\kappa > 0$ such that*

(3.27) $$E[U_\infty^\sigma - U_{T-}^\sigma|\mathcal{F}_T] \leq \kappa E[V_{\sigma-}|\mathcal{F}_T]$$

for all stopping times σ and T, then the ratio inequality

(3.28) $$E[U_\infty^p\exp(\alpha U_\infty/V_\infty)] \leq C E[U_\infty^p]$$

holds for every $0 < p < \infty$ and every α with $0 \leq \alpha < 1/\kappa$, where C is a constant depending only on κ, α and p.

Here U^σ denotes the process $(U_{t \wedge \sigma})$ as usual. Two lemmas are needed for the proof.

Lemma 3. 3. *Let $A = (A_t, \mathcal{F}_t)$ be a right-continuous increasing process satisfying $E[A_\infty - A_{T-}|\mathcal{F}_T] \leq c$ for all stopping times T, with a constant $c > 0$. Then for $0 \leq \alpha < 1/c$ the inequality*

$$E[\exp\{\alpha(A_\infty - A_{T-})\}|\mathcal{F}_T] \leq \frac{1}{1-\alpha c}$$

holds for all stopping times T.

For the proof, see [6].
The second lemma, which was given by T. Murai and A. Uchiyama ([63]), plays an important role in our investigation.

Lemma 3. 4. *Let X and Y be positive random variables on Ω. If there exist two constants $a > 0$ and $c > 0$ such that*

$$(3.29) \qquad\qquad P(X > \gamma\lambda, Y \leq \lambda) \leq ce^{-a\gamma}P(X > \lambda)$$

for every $\lambda > 0$ and every $\gamma > 1$, then the ratio inequality

$$(3.30) \qquad\qquad E[X^p \exp(bX/Y)] \leq C_{b,p}E[X^p]$$

holds for every $p > 0$ and every b with $0 < b < a$, where $C_{b,p}$ is a constant depending only on b and p.

Proof. We shall show this lemma, following the idea of Murai and Uchiyama. Firstly, let us choose a number $\alpha > 1$ satisfying $0 < b < a/\alpha^2$, and let

$$A_{ij} = \{\omega \in \Omega : \alpha^i < X(\omega) \leq \alpha^{i+1}, \alpha^{i-j-1} < Y(\omega) \leq \alpha^{i-j}\}$$

where $i \in \mathbb{Z}$ and $j \in \mathbb{N}$. Then

$$E[X^p \exp(bX/Y)] \leq \sum_{i=-\infty}^{\infty} \sum_{j=1}^{\infty} E[X^p \exp(bX/Y) : A_{ij}] + E[X^p]\exp(b\alpha^2),$$

and so it is enough to estimate the first term on the right-hand side. By an elementary calculation it follows from the assumption (3.29) that

$$\sum_{i=-\infty}^{\infty} \sum_{j=1}^{\infty} E[X^p \exp(bX/Y) : A_{ij}]$$

$$\leq \sum_{i=-\infty}^{\infty} \sum_{j=1}^{\infty} \alpha^{p(i+1)} \exp(b\alpha^{j+2})P(X > \alpha^i, Y \leq \alpha^{i-j})$$

$$\leq c\alpha^p \sum_{i=-\infty}^{\infty} \sum_{j=1}^{\infty} \alpha^{pi} \exp(b\alpha^{j+2}) \exp(-a\alpha^j)P(X > \alpha^{i-j})$$

$$\leq c\alpha^p \sum_{i=-\infty}^{\infty} \sum_{j=1}^{\infty} \alpha^{p(i-j)} P(X > \alpha^{i-j})\alpha^{pj} \exp(b\alpha^{j+2} - a\alpha^j)$$

$$\leq c\alpha^p \sum_{k=-\infty}^{\infty} \alpha^{pk} P(X > \alpha^k) \sum_{j=1}^{\infty} \alpha^{pj} \exp\{(b\alpha^2 - a)\alpha^j\}$$

$$\leq K_{\alpha,p} \sum_{k=-\infty}^{\infty} \alpha^{pk} P(X > \alpha^k)$$

$$\leq C_{\alpha,p}E[X^p],$$

where $K_{\alpha,p} = c\alpha^p \sum_{j=1}^{\infty} \alpha^{pj} \exp\{(b\alpha^2 - a)\alpha^j\}$ and $C_{\alpha,p} = K_{\alpha,p}\alpha^p(\alpha^p - 1)^{-1}$. Thus the lemma is proved. $\square$

Proof of Theorem 3.20 : For each $\lambda > 0$, we first define the two stopping times τ and σ as follows :

$$\tau = \inf\{t : U_t > \lambda\}, \quad \sigma = \inf\{t : V_t > \lambda\}.$$

Obviously, $V_{\sigma-} \leq \lambda$ and so from (3.27) it follows that

$$E[U_\infty^\sigma - U_{T-}^\sigma|\mathcal{F}_T] \leq \kappa\lambda$$

for any stopping time T. Then by Lemma 3.3 we get

$$E\left[\exp\left\{\frac{\delta}{\kappa\lambda}(U_\infty^\sigma - U_{T-}^\sigma)\right\}\middle|\mathcal{F}_T\right] \le \frac{1}{1-\delta}.$$

for every δ with $0 < \delta < 1$. Combining this with the fact that $U_{\tau-} \le \lambda$, it follows that for every $\gamma > 1$

$$
\begin{aligned}
P&(U_\infty > \gamma\lambda, V_\infty \le \lambda) \\
&\le P(U_\infty - U_{\tau-} > (\gamma - 1)\lambda, \tau < \infty, \sigma = \infty) \\
&\le P\left(\frac{\delta}{\kappa\lambda}(U_\infty^\sigma - U_{\tau-}^\sigma) > \frac{\delta(\gamma-1)}{\kappa}, \tau < \infty\right) \\
&\le \exp\left\{-\frac{\delta(\gamma-1)}{\kappa}\right\} E\left[E\left[\exp\left\{\frac{\delta}{\kappa\lambda}(U_\infty^\sigma - U_{\tau-}^\sigma)\right\}\middle|\mathcal{F}_\tau\right] : \tau < \infty\right] \\
&\le C_\delta \exp\left(-\frac{\delta}{\kappa}\gamma\right) P(U_\infty > \lambda),
\end{aligned}
$$

where $C_\delta = \frac{1}{1-\delta}\exp(\frac{\delta}{\kappa})$. Let now $0 \le \alpha < \delta/\kappa$. Then (3.28) follows at once from Lemma 3.4. This completes the proof. □

It is not difficult to verify that any two of the three increasing processes $X^*, \langle X\rangle^{1/2}$ and L^* satisfy the condition (3.27). Thus the following is an immediate consequence of Theorem 3.20.

Theorem 3. 21. *Let U and V be any two of the three increasing processes $X^*, \langle X\rangle$ and L^*. Then for <u>some</u> $\alpha > 0$ the ratio inequality*

$$E[U_\infty^p \exp(\alpha U_\infty/V_\infty)] \le C_{\alpha,p} E[U_\infty^p] \quad (0 < p < \infty)$$

holds for all continuous martingales X.

Furthermore, if $0 \le \beta < 1$, then the inequality

(3.31) $$E[U_\infty^p \exp\{\alpha(U_\infty/V_\infty)^\beta\}] \le C_{\alpha,\beta,p} E[U_\infty^p]$$

is valid for <u>every</u> $\alpha > 0$.

We especially remark the following.

Corollary 3. 7. *For <u>every</u> $0 \le \alpha < \infty$ and <u>every</u> $0 < p < \infty$ the ratio inequalities*

(3.32) $$E[\langle X\rangle_\infty^{p/2} \exp(\alpha\langle X\rangle_\infty^{1/2}/X^*)] \le c_{\alpha,p} E[\langle X\rangle_\infty^{p/2}].$$

(3.33) $$E[\langle X\rangle_\infty^{p/2} \exp(\alpha\langle X\rangle_\infty^{1/2}/L_\infty^*)] \le C_{\alpha,p} E[\langle X\rangle_\infty^{p/2}].$$

holds for all continuous martingales X.

Proof. The usual stopping argument enables us to assume that X is an L^2-bounded martingale. Observe first that $E[\langle X\rangle_\infty - \langle X\rangle_T|\mathcal{F}_T] \le E[(X^*)^2|\mathcal{F}_T]$ for any stopping time T. Then, applying (3.31) to the case where $U = \langle X\rangle, V = (X^*)^2$ and $\beta = 1/2$ we can obtain (3.32). The same argument guarantees the validity of (3.33), because $E[\langle X\rangle_\infty - \langle X\rangle_T|\mathcal{F}_T] \le cE[(L_\infty^*)^2|\mathcal{F}_T]$. Thus the proof is complete. □

It is natural to ask if the similar inequalities for other pairs hold for every $\alpha \geq 0$. But we cannot settle this question so far.

Finally, we make mention of the weighted ratio inequalities. For that, let M be a continuous local martingale, and assume that $\mathcal{E}(M)$ is a uniformly integrable martingale. Then, repeating the same arguments as in Section 3.6 gives the weighted ratio inequality for increasing processes.

Theorem 3. 22.   *Let U and V be two right-continuous increasing processes such that $U_0 = V_0 = 0$. If the martingale M belongs to the class BMO and if there is a constant $\kappa > 0$ such that*

$$E[U_\infty^\sigma - U_{T-}^\sigma | \mathcal{F}_T] \leq \kappa E[V_{\sigma-} | \mathcal{F}_T]$$

for any stopping times σ and T, then the weighted ratio inequality

$$\hat{E}[U_\infty^p \exp(\alpha U_\infty / V_\infty)] \leq C(\kappa, \alpha, p)\hat{E}[U_\infty^p] \quad (0 < p < \infty)$$

holds for some $\alpha > 0$. Moreover, if $0 \leq \beta < 1$, then for every $\alpha \geq 0$

$$\hat{E}[U_\infty^p \exp\{\alpha(U_\infty / V_\infty)^\beta\}] \leq C(\kappa, \alpha, \beta, p)\hat{E}[U_\infty^p] \quad (0 < p < \infty).$$

For the proof, see [42].

As an immediate consequence of this result, we obtain the following.

Corollary 3. 8.   *Let $0 \leq \alpha < \infty$ and $0 < p < \infty$. If $M \in BMO$, then the weighted ratio inequalities*

$$\hat{E}[\langle X \rangle_\infty^p \exp(\alpha \langle X \rangle_\infty^{1/2} / X^*)] \leq C_{\alpha,p}\hat{E}[\langle X \rangle_\infty^p],$$
$$\hat{E}[(X^*)^p \exp(\alpha X^* / \langle X \rangle_\infty^{1/2})] \leq C_{\alpha,p}\hat{E}[\langle X \rangle_\infty^{p/2}],$$
$$\hat{E}[\langle X \rangle_\infty^p \exp(\alpha \langle X \rangle_\infty^{1/2} / L^*)] \leq C_{\alpha,p}\hat{E}[\langle X \rangle_\infty^p]$$

hold for all continuous martingales X.

Bibliography

[1] M. T. Barlow and M. Yor. *(Semi-) Martingale inequalities and local times*, Z. Wahrsch. Verw. Gebiete 55 (1981), 237-254.

[2] N. L. Bassily and J. Mogyoródi. *On the BMO_ϕ-spaces with general Young function*, Annales Univ. Sci. Budapest, Sec. Math. 27 (1984), 225-227.

[3] A. Bonami and D. Lépingle. *Fonction maximale et variation quadratique des martingales en présence d'un poids*, Séminaire de Probabilités XIII, Université de Strasbourg (Lecture Notes in Math. 721, pp.294-306), Berlin Heidelberg New York, Springer 1979.

[4] C.S. Chou and P.A. Meyer. *Sur la représentaiton des martingales comme intégrals stochastiques dans les processus ponctuels*, Séminaire de Probabilités IX, Université de Strasbourg (Lecture Notes in Math. 465,pp.226- 236), Berlin Heidelberg New York, Springer 1975.

[5] R. R. Coifman and C. Fefferman. *Weighted norm inequalities for maximal functions and singular integrals*, Studia Math. 51 (1974), 241-150.

[6] C. Dellacherie and P. A. Meyer. *Probabilités et Potentiel* (second edition) vol.II, Paris, Herman 1980.

[7] C. Dellacherie, P. A. Meyer and M. Yor. *Sur certaines propriétés des espaces de Banach H^1 et BMO*, Séminaire de Probabilités XII, Université de Strasbourg (Lecture Notes in Math. 649,pp.98- 113), Berlin Heidelberg New York, Springer 1978.

[8] C. Doléans-Dade. *Variation quadratique des martingales continues à droite*, Ann. Math. Stat. 40 (1969), 284-289.

[9] C. Doléans-Dade. *Quelques applications de la formule de changement de variables pour les semi-martingales*, Z. Wahrsch. Verw. Gebiete 16 (1970), 181-194.

[10] C. Doléans-Dade and P. A. Meyer. *Une caracterization de BMO*, Séminaire de Probabilités XI, Université de Strasbourg (Lecture Notes in Math. 581, pp.383-389), Berlin Heidelberg New York, Springer 1977.

[11] C. Doléans-Dade and P. A. Meyer. *Inégalités de normes avec poids*, Séminaire de Probabilités XIII, Université de Strasbourg (Lecture Notes in Math. 721, pp.313-331), Berlin Heidelberg New York, Springer 1979.

[12] R. Durrett. *Brownian Motion and Martingales in Analysis*, Belmont, Calif., Wadsworth 1984.

[13] R. J. Elliott. *Stochastic Calculus and Applications*, Berlin Heidelberg New York, Springer 1982.

[14] M. Emery. *Le théorème de Garnett-Jones d'après Varopoulos*, In : J. Azéma, M. Yor (eds:), Séminaire de Probabilités XV, Université de Strasbourg (Lecture Notes in Math. 721, pp.278-284), Berlin Heidelberg New York, Springer 1985.

[15] M. Emery. *Une définition faible de BMO*, Ann. Inst. Henri Poincaré, Prob. Statist. 21 no.1 (1985), 59-71.

[16] M. Emery, C. Stricker and J. A. Yan. *Valeurs prisés par les martingales locales continues à un instant donné*, Ann. Probability 11 (1983), 635- 641.

[17] C. Fefferman. *Characterizations of bounded mean oscillation*, Bull. Amer. Math. Soc. 77 (1971), 587-588.

[18] J. Garnett and P. Jones. *The distance in BMO to L^∞*, Ann. Math. 108 (1978), 373-393.

[19] A. M. Garsia. *Martingale Inequalities*, Seminar Notes on Recent Progress, New York, Benjamin 1973.

[20] F. W. Gehring. *The L^p-integrability of the partial derivatives of a quasi-conformal mapping*, Acta Math. 130 (1973), 265-277.

[21] R. K. Getoor and M. J. Sharpe. *Conformal martingales*, Inventiones math. 16 (1972), 271-308.

[22] I. I. Gihman and A. V. Skorohod. *Stochastic Differential Equations*, Berlin Heidelberg New York, Springer-Verlag 1972.

[23] I. V. Girsanov. *On transforming a certain class of stochastic processes by absolutely continuous substitution of measures*, Theor. Prob. Appl. 5 (1960), 285-301.

[24] R. F. Gundy. *The density of the area integral*(Conference on Harmonic Analysis in Honor of Antoni Zygmund vol. 1, pp. 138-149), Belmont, Calif., Wadsworth 1982.

[25] G. H. Hardy and J. E. Littlewood. *A maximal theorem with function-theoretic applications*, Acta Math. 54 (1930), 81-116.

[26] K. Ito. *Stochastic integral*, Proc. Imperial Acad. Tokyo 20 (1944), 519-524.

[27] K. Ito and H. P. McKeans. *Diffusion Processes and Their Sample Paths*, Berlin, Springer-Verlag 1965.

[28] M. Izumisawa and N. Kazamaki. *Weighted norm inequalities for martingales*, Tôhoku Math. J. 29 (1977), 115-124.

[29] M. Izumisawa, T. Sekiguchi and Y. Shiota. *Remarks on a characterization of BMO-martingales*, Tôhoku Math. J. 31 (1979), 281-284.

[30] T. Jeulin and M. Yor. *Grossissement d'une filtration er semimartingales : formules explicite*, Séminaire de Probabilités XII, Université de Strasbourg (Lecture Notes in Math. 649, pp.78-79), Berlin Heidelberg New York, Springer 1978.

[31] F. John and L. Nirenberg. *On functions of bounded mean oscillation*, Comm. Pure Appl. Math. 14 (1961), 415-426.

[32] N. Kazamaki. *A characterization of BMO-martingales*, Séminaire de Probabilités X, Université de Strasbourg (Lecture Notes in Math. 511, pp.536-538), Berlin Heidelberg New York, Springer 1976.

[33] N. Kazamaki. *On a problem of Girsanov*, Tôhoku Math. J. 29 (1977), 597 -600.

[34] N. Kazamaki. *A property of BMO-martingales*, Math. Rep. Toyama Univ. 1 (1978), 55-63.

[35] N. Kazamaki. *A sufficient condition for the uniform integrability of exponential martingales*, Math. Rep. Toyama Univ. 2 (1979), 1-11.

[36] N. Kazamaki. *An elementary proof of a theorem of Novikov on exponential martingales*, Math. Rep. Toyama Univ. 2 (1979), 65-68.

[37] N. Kazamaki. *Transformation of H^p-martingales by a change of law*, Z. Wahrsch. Verw. Gebiete 46 (1979), 343-349.

[38] N. Kazamaki. *A counterexample related to A_p-weghts in martingale theory*, Séminaire de Probabilités XIX, Université de Strasbourg (Lecture Notes in Math. 1123, pp.275-277), Berlin Heidelberg New York, Springer 1985.

[39] N. Kazamaki. *A new aspect of L^∞ in the space of BMO-martingales*, Probab. Th. Rel. Fields 78 (1988), 113-126.

[40] N. Kazamaki. *Exponential martingales and H^p*, Math. J. Toyama Univ. 14 (1991), 209-211.

[41] N. Kazamaki and M. Kikuchi. *Quelques inégalités des rapports pour martingales continues*, C. R. Acad. Sci. Paris. Sér. 1 305 (1987), 37-38.

[42] N. Kazamaki and M. Kikuchi. *Some remarks on ratio inequalities for continuous martingales*, Studia Math. 94 no.1 (1989), 97-102.

[43] N. Kazamaki and T. Sekiguchi. *On the transforming the class of BMO-martingales by a change of law*, Tôhoku Math. J. 31 (1979), 261-279.

[44] N. Kazamaki and T. Sekiguchi. *Un critère d'intégrabilité uniforme des martingales exponentielles continues*, C. R. Acad. Sci. Paris. Sér. 1 295 (1982), 17.

[45] N. Kazamaki and T. Sekiguchi. *Uniform integrability of continuous exponential martingales*, Tôhoku Math. J. 35 (1983), 289-301.

[46] N. Kazamaki and T. Sekiguchi. *A remark on L^{∞} in the space of BMO-martingales*, Math. Rep. Toyama Univ. 10 (1987), 169-173.

[47] N. Kazamaki and Y. Shiota. *Remarks on the class of continuous martingales with bounded quadratic variation*, Tôhoku Math. J. 37 (1985), 101-106.

[48] M. Kikuchi. *The best estimation of a ratio inequality for continuous martingales*, Séminaire de Probabilités XXIII, Université de Strasbourg (Lecture Notes in Math. 1372, pp.52-56), Berlin Heidelberg New York, Springer 1989.

[49] M. Kikuchi. *Improved ratio inequalities for martingales*, Studia Math. 99 (1991), 109-113.

[50] M. Kikuchi. *A note on the energy inequalities for the increasing processes*, Séminaire de Probabilités XXVI, Université de Strasbourg (Lecture Notes in Math. 1526, pp.533-539), Berlin Heidelberg New York, Springer 1992.

[51] E. Lenglart. *Transformation des martingales locales par changement absolument continu de probabilité*, Z. Wahrsch. Verw. Gebiete 39 (1977), 65-70.

[52] D. Lépingle and J. Mémin. *Intégrabilité uniforme et dans L^r des martingales exponentielles*, Séminaire de Rennes (1978).

[53] D. Lépingle and J. Mémin. *Sur l'intégrabilité uniforme des martingales exponentielles*, Z. Wahrsch. Verw. Gebiete 42 (1978), 175-203.

[54] R. Sh. Liptser and A. V. Shiryayev. *On absolute continiuty of measures associated with processes of diffusion type with respect to the Wiener measure*, Izv. A.N. USSR, Ser. Math. 36 (1972), 847-889(In Russian).

[55] B. Maisonneuve. *Quelques martingales remarquables associées à une martingale continue*, Publ. Inst. Stat. Univ. Paris 3 (1968), 13-27.

[56] H. P. McKean. *Stochastic Integrals*, New York, Academic Press 1969.

[57] M. Métivier. *Semimartingales : a course on stochastic processes*, Berlin New York, de Gruyter 1982.

[58] P. A. Meyer. *Probabilités et Potentiel*, Paris, Hermann 1966.

[59] P. A. Meyer. *Le dual de H^1 est BMO (cas continu)*, Séminaire de Probabilités VII, Université de Strasbourg (Lecture Notes in Math. 321, pp.136-145), Berlin Heidelberg New York, Springer 1973.

[60] P. A. Meyer. *Un cours sur les intégrales stochastiques*, Séminaire de Probabilités X, Université de Strasbourg (Lecture Notes in Math. 511, pp.245-400), Berlin Heidelberg New York, Springer 1976.

[61] P. A. Meyer. *Sur un théorème de C. Herz et D. Lépingle*, Séminaire de Probabilités XI, Université de Strasbourg (Lecture Notes in Math. 581, pp.465-469), Berlin Heidelberg New York, Springer 1977.

[62] B. Muckenhoupt. *Weighted norm inequalities for the Hardy maximal function*, Trans. Amer. Math. Soc. 165 (1972), 207-226.

[63] T. Murai and A. Uchiyama. *Good λ inequalities for the area integral and the nontangential maximal function*, Studia Math. 83 (1986), 251-262.

[64] A. A. Novikov. *On an identity for stochastic integrals*, Theor. Prob. Appl. 17 (1972), 717-720.

[65] A. A. Novikov. *On conditions for uniform integrability of continuous exponential martingales*, (Lecture Notes in Control and Information Science, pp.304-310) Berlin Heidelberg New York, Springer 1978.

[66] T. Okada. *On conditions for uniform integrability of exponential martingales*, Tôhoku Math. J. 34 (1982), 495-498.

[67] I. V. Pavlov. *A counterexample to the conjecture that H^∞ is dense in BMO*, Theor. Prob. Appl. 25 (1980), 152-155.

[68] P. Protter. *An extension of Kazamaki's results on BMO differentials*, Ann. Probab. 8 no.6 (1980), 1107-1118.

[69] P. Protter. *Stochastic Integration and Diffrential Equations. A new approach*, Appl. of Mathematics no.321, Springer-Verlag 1990.

[70] D. Revuz and M. Yor. *Continuous Martingales and Brownian Motion*, Grundlehren der mathematischen Wissenschaften 293, Springer-Verlag 1991.

[71] H. Sato. *Uniform integrabilities of an additive martingale and its exponential*, Stochastics and Stochastics Reports 30 (1990), 163-169.

[72] H. Van Schuppen and E. Wong. *Transformation of local martingales under a change of law*, Ann. Probab. 2 (1974), 879-888.

[73] L. A. Shepp. *Explicit solutions to some problems of optimal stopping*, Ann. Math. Statist. 40 no.3 (1969), 993-1010.

[74] T. Sekiguchi. *BMO-martingales and inequalities*, Tôhoku Math. J. 31 (1979), 355-358.

[75] T. Sekiguchi. *Weighted norm inequalities on the martingale theory*, Math. Rep. Toyama Univ. 3 (1980), 37-100.

[76] T. Sekiguchi and Y. Shiota. *The equivalence of the Muckenhoupt (A_2)-condition and the probabilistic (A_2)-condition*, (unpublished).

[77] E. M. Stein. *On certain operators on L^p spaces*, Doctoral Dissertation III, Chicago, University of Chicago 1955.

[78] J. Stoyanov. *Counterexamples in Probability*, Wiley series in probability and mathematical statistics, Wiley 1987.

[79] D. W. Stroock. *Applications of Fefferman-Stein type interpolation to a probability theory and analysis*, Comm. Pure Appl. Math. 26 (1973), 477-495.

[80] D. W. Stroock and S. R. S. Varadhan. *Diffusion processes with continuous coefficients I*, Comm. Pure Appl. Math. 22 (1969), 345-400.

[81] T. Tsuchikura. *A remark to weighted means of martingales*, Seminar on real analysis, Hachiooji, 1976 (in Japanese).

[82] A. Uchiyama. *Weight functions on probability spaces*, Tóhoku Math. J. 30 (1978), 463-470.

[83] N. Th. Varopoulos. *A probabilistic proof of the Garnett-Jones theorem on BMO*, Proc. J. Math. 90 (1980), 201-221.

[84] S. Watanabe. *A remark on the integrability of* $\sup_t X_t$ *for martingales*, Proc. of the second Japan-USSR symposium on Probability Theory, no.2 (1972), 147-155.

[85] C. Watari. *Calderón-Zygmund's decomposition of martingales and its applications*, Seminar on real analysis, Hachiooji, 1976 (in Japanese).

[86] J. Yan. *Criteria for the uniform integrability of exponential martingales*, Acta Mathematica Sinica 23 (1980), 293-300.

[87] J. Yan. *Sur un théoreme de Kazamaki-Sekiguchi*, Séminaire de Probabilités XXII, Université de Strasbourg (Lecture Notes in Math. 986, pp.121-122), Berlin Heidelberg New York, Springer 1983.

[88] M. Yor. *Application de la relation de domination à certains renforcements des inégalités de martingales*, Séminaire de Probabilités XVI, Université de Strasbourg (Lecture Notes in Math. 920, pp.221-233), Berlin Heidelberg New York, Springer 1982.

Index

Lecture Notes in Mathematics

For information about Vols. 1–1394
please contact your bookseller or Springer-Verlag